MATHEMATICAL TEXTS

FOR COLLEGES

EDITED BY

PERCEY F. SMITH, Ph.D.

PROFESSOR OF MATHEMATICS IN THE SHEFFIELD SCIENTIFIC SCHOOL OF YALE UNIVERSITY

PROBLEMS IN THE CALCULUS

WITH FORMULAS AND SUGGESTIONS

BY

DAVID D. LEIB, Ph.D.

INSTRUCTOR IN MATHEMATICS IN THE SHEFFIELD SCIENTIFIC SCHOOL
OF YALE UNIVERSITY

GINN AND COMPANY

BOSTON · NEW YORK · CHICAGO · LONDON
ATLANTA · DALLAS · COLUMBUS · SAN FRANCISCO

5

215.10

The Athenæum Press
GINN AND COMPANY · PROPRIETORS · BOSTON · U.S.A.

PREFACE

The present volume is the outgrowth of lists of problems prepared by the author from year to year to supplement the textbook used in the Sheffield Scientific School. Many of the problems have been furnished by his colleagues through the media of test and examination papers, or by direct contribution to this collection at some stage in its development. Since many of these in turn were doubtless adapted from other sources, no attempt has been made to assign a problem to an original source. The principles embodied in the problems are surely common property. The book does not aim to be a textbook on the calculus, nor simply a collection of applied problems in science and engineering. It is believed that a teacher can find here a supplementary list of workable problems on any topic ordinarily included in a general course in the calculus. No attempt is made to explain the theory of any science involved in a problem except in so far as it is necessary to an intelligent understanding of the problem and its purpose. The text introducing the exercises aims to explain the technique of the subject and to point out some common pitfalls to the student. The answers to a large number of the problems have been purposely omitted. The general object has been to give the answer to one or more examples of each type so that the student may attack further examples of a similar nature with increased confidence. At the same time, other answers have been omitted so that the book may be used in tests and in work where it is not desirable for the student to have the answers. Comparatively few illustrative examples have been worked out in detail, and practically no figures have been inserted, as these can be found in sufficient quantity in the textbook used. The drawing of suitable figures where needed is a valuable part of an example, and should be insisted upon. The book will be found to differ materially from the problem books, mostly in German, now available. A good bibliography of these is to be found in the *Bulletin of the American Mathematical Society*, June, 1914.

The author desires to thank in particular his colleagues, Drs. Harris F. MacNeish and George F. Gundelfinger, for extensive aid — the former in the differential calculus, the latter in the integral. His thanks are also due to Captain James Gordon Steese, U.S.A., for several valuable suggestions. As no one has formally verified the answers nor read the proof, the author must assume sole responsibility for errors and omissions, which he hopes will not be sufficiently numerous to affect the value of the book. The author will be pleased to have his attention called to errors either in the text or in the answers.

DAVID D. LEIB

SHEFFIELD SCIENTIFIC SCHOOL
YALE UNIVERSITY

CONTENTS

CHAPTER I

FUNCTIONS

CHAPTER II

FORMAL DIFFERENTIATION

CHAPTER III

SIMPLE APPLICATIONS. SUCCESSIVE DIFFERENTIATION

CHAPTER IV

MAXIMA AND MINIMA

CHAPTER V

DIFFERENTIALS AND RATES

CHAPTER VI

INDETERMINATE FORMS

CHAPTER VII

CURVATURE. EVOLUTES

CHAPTER VIII

PARTIAL DERIVATIVES. APPLICATIONS

CHAPTER IX

SERIES

CHAPTER X

FURTHER APPLICATIONS TO GEOMETRY

CHAPTER XI

INTEGRAL CALCULUS. SIMPLE FORMAL INTEGRATION

CHAPTER XII

ELEMENTARY APPLICATIONS. DEFINITE INTEGRALS

CHAPTER XIII

SPECIAL METHODS OF INTEGRATION. APPLICATIONS

CHAPTER XIV

INTEGRATION A SUMMATION PROCESS. GEOMETRICAL APPLICATIONS

CHAPTER XV

APPROXIMATE INTEGRATION — VARIOUS METHODS

CHAPTER XVI

MULTIPLE INTEGRATION

CHAPTER XVII

GENERAL APPLICATIONS

CHAPTER XVIII

DIFFERENTIAL EQUATIONS

PROBLEMS IN THE CALCULUS

CHAPTER I

FUNCTIONS

EXERCISE I

Evaluation of Functions. The idea of a function is fundamental. Remember that if $f(x)$ is any expression involving x, whether algebraic or transcendental, $f(a)$ is the same expression where each x has been replaced by a. The following examples should fix this idea:

1. Given $f(x) = x^3 - 2x^2 - 13x + 10$, show that $f(1) = f(2)$.

2. Given $f(x) = x^3 - 5x^2 + 8x - 4$, show that $f(2) = f(1)$.

3. Given $f(x) = x^3 - 3x^2 + 4x - 1$, show that $f(2) = 3f(1)$.

4. Given $f(x) = x^3 + 4x^2 - 5x + 7$, show that $f(2) = 3f(1)$.

5. Given $f(x) = 4x^3 - 3x^2 - 5x - 6$, show that $f(3) + 6f(1) = 0$.

6. Given $f(x) = 2x^3 - 2x^2 - 3x - 3$, show that $f(-3) = 3f(-1)$.

7. Given $f(x) = x^3 - 6x^2 + 11x - 10$, show that $f(1) = f(2) = f(3)$.

8. Given $\phi(x) = x^3 - 9x^2 + 26x - 14$, show that $\phi(2) = \phi(3) = \phi(4) = \frac{1}{4}\phi(5)$.

9. Given $f(x) = x + 1/x$, show that $f(a) = f(1/a)$.

10. Given $f(x) = \sin x + \cos x$, show that $f(\pi/3)f(2\pi/3) = 1/2$.

11. Given $\phi(x) = \tan x + \tan 3x$, show that $\phi(\pi/3) + \phi(2\pi/3) = 0$.

12. Given $\psi(x) = \sin x/(1 + \cos x)$, show that $\psi(\pi/3) \cdot \psi(2\pi/3) = 1$.

Evaluate the following for the given values of the independent variable.

13. $f(x) = x + \log_{10} x$; $x = 1, 2, 3, 1/10, 7, 100$.

14. $f(x) = 64^{1/x}$; $x = 1, 2, 3, 4, 11, -2$.

15. $f(t)=5(e^t+e^{-t})$; $t=1/2, 1, 2$.

16. $f(x)=(e^x+e^{-x})^{x+1}$, $x=0, 1$.

17. $f(x)=\log(x^2-4x)^2$, $x=2, 3$.

18. $f(t)=4e^{-t/4}\sin 2t$,
$t=1, 2, 10$.

19. $f(t)=10e^{t/2}\sin t$, $t=1, 2, 4$.

20. $f(x)=(1+\log x)^{x^2}$,
$x=1, 2$.

21. $f(t)=e^{-t/5}(\sin t+2\cos t)$,
$t=0, \pm 1, \pm 5$.

22. $f(x)=x^2\cos\frac{1}{3}\pi(x+1)$,
$x=0, \pm 1, \pm 2$.

23. $f(x)=2^{x^2}\sin(1-x)$,
$x=0, 1, 2$.

24. $f(t)=t\sin(t+\pi/4)$,
$t=0, 1, 2$.

In the remaining examples, show the required functional relation.

25. Given $f(x)=\sin x$, show $f(2x)=2f(x)\sqrt{1-f^2(x)}$.

26. Given $\psi(x)=\log x$, show $\psi(ab)=\psi(a)+\psi(b)$.

27. Given $\phi(x)=\sin x$, show $\phi(x)+\phi(y)=2\phi[(x+y)/2]\sqrt{1-\phi^2[(x-y)/2]}$.

28. Given $f(x)=\cos x$, show $f(x)+f(y)=2f[(x+y)/2]f[(x-y)/2]$.

29. Given $f(x)=a^x$, show $f(x+y)=f(x)f(y)$ and $f(x-y)=f(x)/f(y)$.

30. Given $f(x)=1/x$, show $[f(x)+f(y)]\div[f(x)-f(y)]=-(x+y)\div(x-y)$.

31. Given $f(x)=1+1/x$, show $2+f(x)f(y)=f(xy)+f(x)+f(y)$.

32. Given $\psi(x)=x+1/x$, show $\psi(x^2)=\psi^2(x)-2$.

33. Given $\phi(x)=\log(x+1)$, show $\phi(y^2-2)-\phi(y-2)=\phi(y)$.

34. Given $\phi(x)=\log 1/x$, show $\phi(y-1)-\phi(y^2-1)=\phi[1/(y+1)]$.

35. Given $\phi(x)=e^{x-1}$, show $\phi(x^2)=\phi(2x^2-x)\phi(1+x-x^2)$.

36. Given $\phi(x)=a^{x+2}$, show $\phi(y-2)\phi(4-y)=\phi(4)$.

37. Given $f(x)=a^{2-3x}$, show $f(y-1)f(2-y)=f(1/3)$.

38. Given $f(x)=e^{2/x}$, show $f(y+1)f(y-1)=f[(y^2-1)/2y]$.

EXERCISE II

Indeterminate Forms. When a function assumes for a given value of the variable an indeterminate form (such as $0/0$, ∞/∞, etc.), it is possible in many cases to find the limiting value by elementary means. If both numerator and denominator vanish for $x=a$, then $x-a$ may be taken out as factor in both numerator and denominator, when these are polynomials. The student should observe, in the same case, how the coefficients of the highest powers of x determine the limiting value for $x = \infty$, and how the coefficients of the terms of lowest degree determine the limiting value for $x = 0$. Calculate the limiting values of the function for the given values of x.

1. $\dfrac{x^2 - 4x + 3}{x^2 - 5x + 4}$, $x = 1$.

2. $\dfrac{x}{3x - x^2}$, $x = 0$.

3. $\dfrac{x^2 + x}{3x^2 - 2x}$, $x = 0$.

4. $\dfrac{x + 3}{2x - 5}$, $x = \infty$.

5. $\dfrac{x + 3}{2x^2 - 5}$, $x = \infty$.

6. $\dfrac{x^2 + 3}{2x - 5}$, $x = \infty$.

7. $\dfrac{3x^3 - x^2 - 10}{5x^3 + x + 4}$, $x = \infty$.

8. $\dfrac{3x^3 - 2x^2 + 5x}{x^2 - x}$, $x = 0$.

9. $\dfrac{x^2 + 2x + 1}{x^3 - 5x^2 + 7}$, $x = \infty$.

10. $\dfrac{3x^2 - 2x + 5}{x^2 - 4x + 7}$, $x = \infty$.

11. $\dfrac{(n+1)(n+2)(n+3)}{(n-1)(n-2)(n-3)}$, $n = \infty$.

12. $\dfrac{t^3 - 3t^2 + t}{t^3 + 5t^2 + 2t}$, $t = 0, \infty$.

13. $\dfrac{t^4 - 6t + 7}{2t^4 + t^3 + t^2}$, $t = \infty$.

14. $\dfrac{x^3 - 6x}{x^3 - 2x^2 + 3x}$, $x = 0, \infty$.

15. $\dfrac{\sin x}{\tan x}$, $x = 0$.

16. $\dfrac{\tan x + \sec x - 1}{\tan x - \sec x + 1}$, $x = 0$.

17. $\dfrac{\sin \theta}{\sin 2\theta}$, $\theta = 0$.

18. $\dfrac{\tan x}{\tan 3x}$, $x = \dfrac{\pi}{2}$.

CHAPTER II

FORMAL DIFFERENTIATION

EXERCISE III

Differentiation by the General Method. In the following, x is regarded as the independent variable, and y is a function of x. Differentiation is then accomplished as follows: Replace x on the right side by $x + \Delta x$ (or give x an increment Δx) and call the resulting value of the function $y + \Delta y$. Then, by subtraction, we have the value of Δy (the increment of y) expressed in terms of x and Δx. *This expression should be simplified as much as possible.* Next divide both sides by Δx, getting a value for $\Delta y/\Delta x$. Then let Δx approach zero, and the left-hand member becomes the derivative, dy/dx, while the right-hand member reduces to a function of x, which may be a mere constant. Differentiate the following by this method.

1. $y = 3x + 7.$
2. $y = 2x - 3.$
3. $x = t^2 + 2.$
4. $y = 2t^2 - 3.$
5. $y = 2t^3 + 1.$
6. $y = 2t^2 - t.$
7. $s = 3t^2 + 2t.$
8. $y = x^2 + 7x.$
9. $y = 2x^3 - 4x.$
10. $y = 2t - t^3.$
11. $y = 2t^2 - t^3.$
12. $y = t^3 - 5t.$
13. $y = 1/x.$
14. $s = 1/3t.$
15. $y = 1/x^2.$
16. $y = 3/(x + 1).$
17. $y = 2/(1 - t).$
18. $y = 3/(2x - 1).$
19. $y = 3/(x^2 - 1).$
20. $x = 5/(2t + 3).$
21. $y = 4/(3 - 2x).$
22. $y = 3/(5 - 4x).$
23. $y = 2/(1 - t^2).$
24. $s = 3/(1 + t^2).$
25. $y = (5 - 2x)/(3 - x).$
26. $y = (2x + 1)/x^2.$
27. $y = x^2/(2x + 1).$
28. $y = x^2/(1 - 3x).$
29. $s = (5t^2 + 1)/2t.$
30. $x = 2y^2/(1 + 2y).$

The value of dy/dx at a point on a curve gives the slope of the tangent to the curve at that point. This enables us to find the angle between two curves at the point of intersection, by the formula $\tan\theta = (m_1 - m_2)/(1 + m_1 m_2)$, where the m's are the slopes of the curves at the point. Find the angle of intersection of the following pairs of curves:

31. $y = 2 - x^2$, $y = x^2$.

32. $y = 5 - x^2$, $y = 5 + 3x$.

33. $y = 3x - 5$, $y = x^2 - 5$.

34. $y = x^2 - 2x + 1$, $y = 7 + 2x - x^2$.

35. $y = x^2 - 4$, $y = 4 - x^2$.

36. $y = x^3 - 2x^2$, $y = 2x - x^2$.

37. $y = x^2 + 2x + 1$, $y = 9 + 2x - x^2$.

38. $y = 1/(1 + x)$, $y = 1/(1 - x)$.

39. $y = 2/(x - 1)$, $y = x^3/4$, at (2, 2).

40. $y = x/(x - 1)$, $y = x^2/2$, at (2, 2).

41. $y = x^3/4$, $y = 6 - x^2$, at (2, 2).

42. $y = \frac{1}{2}(x + 1)$, $y = \frac{1}{4}(x^2 + 2x + 1)$.

EXERCISE IV

Simple Polynomials for Oral Differentiation. Differentiate the expressions below by means of the formulas:

$$\frac{d}{dx}x^n = nx^{n-1}; \qquad \frac{d}{dx}v^n = nv^{n-1}\frac{dv}{dx}.$$

1. $y = x^2 - 2x + 3$.

2. $y = 2x^2 + x + 1$.

3. $y = 3x - x^2$.

4. $y = 2x^3 - 7x + 4$.

5. $y = 7 - x - 2x^2$.

6. $y = 2x^3 + 5x^2 - 2$.

7. $y = 3x^2 - 2x + 1$.

8. $y = \frac{1}{3}x^3 + 3x$.

9. $y = \frac{1}{2}x^2 + 2x - 4$.

10. $y = x^4 - 2x^2 - 3$.

11. $y = x^3 + \frac{1}{2}x^2 - 4$.

12. $y = \frac{1}{2}x^4 + x - 2$.

13. $y = \frac{1}{3}x^3 + 2x^2 - 5$.

14. $y = 3x^2 - \frac{1}{2}x^4$.

15. $y = x^3 + x^2 - x$.

16. $y = \frac{1}{3}x^3 - \frac{1}{2}x^2$.

17. $y = x - \frac{1}{3}x^3 + x^2$.

18. $y = \frac{1}{5}x^5 + 3x^3$.

19. $y = (x + 1)^3$.

20. $y = (x - 1)^2$.

21. $y = (1 - x)^2$.

22. $y = (1 + 2x)^3$.

23. $y = (x^2 - 1)^2$.

24. $y = (x + 6)^7$.

25. $y = (4 - x^2)^3$.

26. $y = (x^3 + 1)^2$.

27. $y = (1 - 2x)^4$.

28. $y = (1 - 3x^2)^2$.

29. $y = (7 - 3x)^4$.

30. $y = (6 - x^2)^2$.

31. $y = (x^2 + 2x - 1)^2$.

32. $y = (x^3 - 2x)^2$.

33. $y = (x^3 - 1)^4$.

34. $y = (4 - 2x^3)^3$.

35. $y = (2 - x^4)^2$.

36. $y = (2x - x^3)^4$.

EXERCISE V

The Power Formula. Fractional and Negative Exponents. The formula $dv^n/dx = nv^{n-1}\,dv/dx$ holds for fractional and negative exponents as well as for positive ones. In differentiating a radical expression, always rewrite with fractional exponents. Similarly, a constant over a power, for example, $4/(x+1)^3$, is most conveniently differentiated by rewriting in the form with a negative exponent; that is, $4/(x+1)^3 = 4(x+1)^{-3}$. Differentiate the following:

1. $y = 2x^{\frac{1}{2}} - 3x^{\frac{2}{3}}$.
2. $y = 2x^{\frac{3}{2}} + \sqrt{2x}$.
3. $y = 1/x + x^{\frac{3}{4}}$.
4. $s = 3t^{\frac{1}{3}} - t^{\frac{5}{2}}$.
5. $r = t^2 + 2\sqrt{t} + 3t^{\frac{4}{3}}$.
6. $y = x^3 + 1/3x^{\frac{1}{3}}$.
7. $y = 2x^3 + 3\sqrt{x} - 3/2x^2$.
8. $y = 3x^{\frac{8}{3}} - x^{\frac{5}{4}}$.
9. $y = \sqrt{2x} + 2\sqrt[3]{x}$.
10. $y = 3/\sqrt{x} + 3\sqrt{x}$.
11. $y = \sqrt{3x} + \sqrt{x}/3$.
12. $y = 7/x^2 - 3x^{\frac{1}{3}} - 1/\sqrt{x}$.
13. $R = \theta + 1/\theta - 2\sqrt{\theta}$.
14. $s = 4/\sqrt{t} + 3/\sqrt[3]{t}$.
15. $s = 4\sqrt{x^3} + 2/3x^3$.
16. $y = 4/(x-1)^2$.
17. $y = \sqrt{x^2 - 4}$.
18. $y = \sqrt[3]{1 - x^3}$.
19. $y = 3/(1-x)^3$.
20. $y = \sqrt{4 - 3x^2}$.
21. $y = 3/\sqrt{x^2 - 2}$.
22. $y = 3/(2x - 5)^2$.
23. $y = 4(x^2 - 2x)^{\frac{1}{2}}$.
24. $s = \sqrt{1 - 2t^3}$.
25. $R = 2\sqrt[3]{(1 + t^2)^2}$.
26. $x = 5/(3 - 2t)^2$.
27. $x = 5/\sqrt[3]{3t^2 - 6t}$.
28. $x = 4/\sqrt{6t - t^2}$.
29. $x = 3\sqrt[5]{5t^2 - t^5}$.
30. $y = \sqrt{4x^3 - 3x^2 + 6x}$.
31. $y = 2\sqrt{(1 + 4x^2)^3}$.
32. $y = 3\sqrt{(5 - 6x^2)^3}$.

EXERCISE VI

Products and Quotients of Algebraic Forms. The formulas are

$$\frac{d}{dx}(uv) = v\frac{du}{dx} + u\frac{dv}{dx}, \qquad \frac{d}{dx}\frac{u}{v} = \left(v\frac{du}{dx} - u\frac{dv}{dx}\right)/v^2.$$

It should be borne in mind that u and v are both functions of x. If either of them is a mere constant, it is obviously not necessary to use these formulas. If the product is merely the product of two simple polynomials, it is preferable to multiply out into a simple polynomial before differentiating. If, in the case of a quotient, both u and v are polynomials, and u is of higher degree than v, it is possible to divide out before differentiating; for example, $(x^3+1)/(x-1) = x^2 + x + 1 + 2/(x-1)$. Apply the above formulas in differentiating the following:

1. $y = (x+2)\sqrt{x^2 + 4x}$.
2. $y = (2x+5)\sqrt{x^2 - 3}$.
3. $y = (2 - 3t)\sqrt{1 - 2t^2}$.
4. $x = (3t - 2)\sqrt{2t^3 - 4t}$.
5. $y = (x - 4)\sqrt{1 - 2x + 4x^2}$.
6. $y = (2x + 3)\sqrt{x^2 + 4x}$.
7. $y = (2x - 1)\sqrt{2x - x^2}$.
8. $y = (2 - t)\sqrt{3t^2 + 2t + 1}$.

9. $y = (x^2 + 2x)\sqrt{2x+3}$.

10. $y = \dfrac{2x+3}{1-x^2}$.

11. $y = \dfrac{2x^2+3x}{1+2x}$.

12. $y = \dfrac{t^2-3t-1}{3t-1}$.

13. $y = \dfrac{t^3+4}{1+2t}$.

14. $y = \dfrac{3-x^3}{2x^2+5}$.

15. $s = \dfrac{2t^2-4t}{3-2t^2}$.

16. $y = \dfrac{x^3-1}{x^2-3x}$.

17. $y = \dfrac{x}{\sqrt{x^2-5}}$.

18. $y = \dfrac{\sqrt{x-1}}{x^2}$.

19. $y = \dfrac{x^2-x}{\sqrt{1-x^2}}$.

20. $y = \dfrac{\sqrt{3-t}}{t^3}$.

21. $y = \dfrac{3-x}{\sqrt{x^2-6x}}$.

22. $y = \dfrac{\sqrt{1+2t^2}}{2t+3}$.

23. $y = \dfrac{\sqrt{1-3x^2}}{x^2}$.

24. $y = \dfrac{\sqrt{3t^2+5}}{1-t}$.

25. $y = \sqrt{1-2x}\ \sqrt{2x^2-3}$.

26. $y = \sqrt[3]{1+3x}\ \sqrt{1+2x}$.

27. $y = \sqrt[3]{1+x^3}\ \sqrt{1-x^2}$.

28. $y = \sqrt{1+x^2}\ \sqrt[3]{1-x^3}$.

29. $y = \sqrt{x^2-2x}\ \sqrt[3]{x^3-3x}$.

30. $y = \sqrt[3]{3x+x^3}\ \sqrt{x^2-4}$.

EXERCISE VII

Review of Algebraic Forms. Differentiate the following:

1. $y = x^3 - 2x^2 - 2/\sqrt{x}$.
2. $y = (x^{\frac{3}{2}} - x^{\frac{1}{2}} + 2)/x^{\frac{1}{2}}$.
3. $y = 3/\sqrt{x} + x/\sqrt{3} + \sqrt{x}/\sqrt{3}$.
4. $y = 3/\sqrt[3]{x} + 2/x^3$.
5. $y = \sqrt{2x^3 + 4x - 5}$.
6. $y = \sqrt{3x + 1/x}$.
7. $y = \sqrt{2t-7} + \sqrt{7-3t}$.
8. $y = (x-1)\sqrt{x^2-2x+2}$.
9. $y = (3x+2)\sqrt{1-2x-x^2}$.
10. $y = (3-2x)\sqrt[3]{x^3-3x}$.
11. $y = (x+x^2)\sqrt[3]{1-x^3}$.
12. $y = (3-x^3)/(2x^2+5)$.
13. $y = (3-x)/\sqrt{x^2-6x}$.
14. $y = \sqrt{1+2t^2}/(2t+3)$.
15. $y = 6t\sqrt{t-2} - 4(t-2)^{\frac{3}{2}}$.
16. $y = x\sqrt{2x+4} - \frac{1}{3}(2x+4)^{\frac{3}{2}}$.
17. $y = 2t\sqrt{1-4t} + \frac{1}{6}(1-4t)^{\frac{3}{2}}$.
18. $y = (2t/\sqrt{1+t}) - 4\sqrt{1+t}$.
19. $y = (x^2/\sqrt{1+2x}) + x\sqrt{1+2x}$.
20. $y = (\sqrt{1+2x}/2x) - 1/\sqrt{1+2x}$.
21. $y = (2x/\sqrt{x+7}) + x\sqrt{x+7}$.
22. $y = x^2\sqrt[3]{3x^2-6}$.
23. $y = [3/(x-7)] + [(x+7)/3x]$.
24. $y = [(2x+3)/(x^2+4)]^2$.
25. $y = [(x^2+5)/(2x+3)]^2$.
26. $y = \sqrt{(3t+1)/(3t-1)}$.
27. $y = (\sqrt{x+5}/2x) + 2x/\sqrt{x+5}$.
28. $y = (2\sqrt{x} + 3/x)/(2\sqrt{x} - 3/x)$.
29. $y = 4/\sqrt[3]{3x^2-6x+5}$.
30. $y = (7+t)\sqrt{3-2t} + t^2$.

EXERCISE VIII

Logarithms of Algebraic Forms, Base *e*. In differentiating logarithmic functions, the forms should always be rewritten when it is possible to simplify the differentiation by so doing. For this purpose it is well to recall the following formulas used in logarithmic work generally:

(a) $\log uv = \log u + \log v$, (c) $\log v^n = n \log v$,

(b) $\log u/v = \log u - \log v$, (d) $\log \sqrt[n]{v} = 1/n \cdot \log v$.

Thus $\log \sqrt[3]{x^3 + 3x^2} = \frac{1}{3} \log (x^3 + 3x^2)$, and

$$\log (x^3 + 3x^2)/\sqrt{x^2 + 6x} = \log (x^3 + 3x^2) - \tfrac{1}{2} \log (x^2 + 6x).$$

This simplification is important. The formula for differentiating the logarithm of any function is

$$\frac{d}{dx} \log v = \frac{1}{v} \frac{dv}{dx}.$$

This is easily remembered by the fact that the derivative is clearly a fraction with v as its denominator and dv/dx as its numerator. The first list is for oral differentiation.

1. $y = \log (x^2 + 1)$.
2. $y = \log (3 + x^2)$.
3. $y = \log (x^3 + 3x)$.
4. $y = \log (3x^2 + 7)$.
5. $y = \log \sqrt{2x^2 + 5}$.
6. $y = \log (x^2 + 2)^2$.
7. $y = 2 \log (x^2 - 4x)$.
8. $y = \log \sqrt[3]{6 - x^3}$.
9. $y = \log 1/x$.
10. $y = \log (x + 5)^4$.
11. $y = 3 \log \sqrt[4]{1 - 2x^2}$.
12. $y = \log (4 - 3x)^3$.
13. $y = \log (2x + 3)^{\frac{1}{2}}$.
14. $y = \log 1/(x - 3)$.
15. $y = \log 1/(2 - 3x)$.
16. $y = \log \sqrt[3]{x}$.
17. $y = \log \sqrt[5]{x^3}$.
18. $y = \log \sqrt[3]{5 + 6x}$.
19. $y = \log \sqrt[3]{3x^2 - 6x}$.
20. $y = \log \sqrt{(x^2 - 4)^3}$.
21. $y = \log x^2 \sqrt{1 - 2x}$.

Differentiate the following:

1. $y = \log (x^2 + 4x + 6)^2$.
2. $y = \log (6x^2 + 2x + 5)^3$.
3. $y = \log (x^3 - 3x^2 + 6x)^{\frac{1}{3}}$.
4. $y = \log 3 \sqrt{1 - x^2}$.
5. $y = \log [(x^2 + 2)/(1 - x)]$.
6. $y = \log (2x^2 - 4x + 3)^{\frac{3}{4}}$.
7. $y = \log [(x^3 - 3x)/(x^2 - 4)]$.
8. $y = \log 3 (t^2 + 5t - 6)^2$.
9. $y = \log (2t^3 - 5t^2 - 4t + 1)^{\frac{1}{2}}$.
10. $y = \log (x^2 + 5x + 3)(x^2 + 4x - 1)$.

11. $y = \log (t^3 + 3t)^{\frac{1}{3}} (t^2 + 1)^{\frac{1}{2}}$.

12. $y = 3 \log [(x^2 + 7x - 3)/(x^2 - 2x + 5)]$.

13. $y = \log [(3x^2 - 5x)/(3x + 1)]$.

14. $y = \log 4 (t^3 + 4t)^3$.

15. $y = \log [(3t^2 - 6t + 4)/(t^2 - 2t + 1)]$.

16. $y = \log (\sqrt[3]{1 + x^3}/\sqrt{1 + x^2})$.

17. $y = \log\left[\sqrt{(3 + x^2)^3}/\sqrt[3]{(3 - x^2)^2}\right]$.

18. $y = \log\left(x + \sqrt{1 + x^2}\right)$.

19. $y = \log\left(2x + \sqrt{4x^2 + 9}\right)$.

20. $y = \log\left(3x + \sqrt{9x^2 - 1}\right)^{\frac{1}{3}}$.

21. $y = 2 \log\left(3x + \sqrt{9x^3 + 16}\right)$.

22. $y = \log\left(2x^2 + \sqrt{4x^4 + 9}\right)$.

23. $y = \log\left(x^3 + \sqrt{x^6 + 4}\right)$.

24. $y = \log \sqrt{x^2 + 4}\, \sqrt[3]{x^3 + 6x}$.

25. $y = \log\left[(1 + x)/x^2 \sqrt[3]{1 - x}\right]$.

26. $y = 2 \log x^3 \sqrt{2x^2 + 6x}$.

27. $y = \log\left(\sqrt[3]{2x^2 - 1}/x\sqrt{1 + x^2}\right)$.

28. $y = \log\left(\sqrt{x} - 1/x\right)$.

29. $y = \log\left[\left(\sqrt{x^2 + 1} - x\right)/\left(\sqrt{x^2 + 1} + x\right)\right]$.

30. $y = \log\left[\left(x^2\sqrt{x^2 - 1}\right)/(x^2 + 1)\right]$.

EXERCISE IX

Logarithms of Algebraic Functions. When the base is changed from e to a, the formula of the preceding exercise is simply multiplied by $\log_a e$, or

$$\frac{d}{dx}\log_a v = \log_a e \frac{1}{v}\frac{dv}{dx}.$$

As before, it is essential that the function be first rewritten in the best form for differentiation. For example, $\log_4 \sqrt{1 + x^2}/x^2$ should be rewritten as $\frac{1}{2}\log_4 (1 + x^2) - 2\log_4 x$. Differentiate the following:

1. $y = \log_7 \sqrt{1 - 2x^3}$.

2. $y = \log_{10} \sqrt[3]{3x + 7}$.

3. $y = \log_4 (2x + \sqrt{6x})$.

4. $y = \log_4 [(2x - 6)/(5 - 6x)]$.

5. $y = \log_3 (x + 1/x)$.

6. $f(x) = \log_a x\sqrt{1 + x^2}$.

7. $f(t) = \log_5 [(t^2 + 1)/(t^2 - 1)]$.

8. $y = \log_2 \sqrt{(1 + x^2)/(1 - x^2)}$.

9. $y = \log_2 x^2\sqrt{1 - x^2}$.

10. $y = \log_6 (1 - 2t)\sqrt[3]{1 - t^3}$.

11. $y = \log_4 \left(x/\sqrt{1 - x^2}\right)$.

12. $s = \log_8 \sqrt{4x - x^2}$.

13. $y = \log_5 (x + \sqrt{1 + x^2})$.

14. $y = \log_5 [t/(1 + t^2)]$.

15. $y = \log_{10} [(1+t)/(t\sqrt{1-t^2})]$.

16. $y = \log_5 (\sqrt{1 + x^2}/\sqrt[3]{1 - x^3})$.

17. $y = \log_7 [(x^2 - 2)/(6 - 2x^2)^{\frac{3}{2}}]$.

18. $y = \log_8 [(1 + \sqrt{x})/(1 - \sqrt{x})]$.

19. $y = \log_6 [(2 + 3x)/(x^2 \sqrt[3]{1 - x^2})]$.

20. $y = \log_3 (2 + x + \sqrt{4x + x^2})$.

21. $y = \log_2 (2x + 3 + \sqrt{4x^2 + 12x + 5})$.

22. $y = \log_3 (2x - 1 + \sqrt{4x^2 - 4x + 5})$.

23. $y = \log_4 (\sqrt{2}x + \sqrt{5 + 2x^2})$.

24. $y = \log_3 (x + 1 + \sqrt{x^2 + 2x})$.

25. $y = \log_4 [(6x + 1)/2 + \sqrt{9x^2 + 3x}]$.

26. $y = \log_4 (1/x^2 - \sqrt{x})$.

27. $y = \log_4 [(\sqrt{1 + x} - \sqrt{1 - x})/(\sqrt{1 + x} + \sqrt{1 - x})]$.

28. $y = \log_{10} [\sqrt[3]{(x^3 - 5)^2}/\sqrt{(5 - x^2)^3}]$.

29. $y = \log_3 [x^3 (x - 1)/(x + 2)^2]$.

30. $y = \log_2 [(3x - 1)^3/x(3x + 1)]$.

EXERCISE X

Powers of Logarithms. This list is designed principally to review the previous formulas and not to introduce any new ones. The student should note carefully the difference between $\log^2 x$ * and $\log x^2$. The latter equals $2 \log x$, but not the former. To differentiate $\log^n v = (\log v)^n$, we must use first of all the power formula. Differentiate the following:

1. $y = \log^2 (2x^3 - 4x)$.

2. $y = \sqrt{\log (x^2 + 6x)}$.

3. $y = \log^3 (x^2 + 7x)$.

4. $y = \log^2 \sqrt{x^4 + 3x^2}$.

5. $y = \log^5 \sqrt{1 - 3x^2}$.

6. $y = 3\sqrt{\log (1 - 2x^2)}$.

7. $y = \log^3 \sqrt[3]{4 - x^3}$.

8. $y = 2\sqrt[3]{\log (2x^3 - 5)}$.

9. $y = \log^4 (2x^2 - 5x^{\frac{1}{2}})$.

10. $y = \log_3^5 \sqrt{1 - 5x^2}$.

11. $y = \log^2 (x^4 - 3x^2 + 5)^3$.

12. $y = 2\log^2 [(x - 7)/(x - 2)]$.

13. $y = \log_2^3 (x^2 - 2/x)$.

14. $y = \log^4 (2x^3 - 1/x^2)$.

15. $y = \log^3 [(3 - 2x)/(4 + 3x)]$.

16. $y = \log^3 (x^3 - 2x^2 + 5x)^2$.

17. $y = \log^3 [(1 - x)/(1 + x)]$.

18. $y = \log^2 [(t^2 - 4)/(t^2 + 4)]^{\frac{1}{2}}$.

19. $y = \log^7 x^2 \sqrt[3]{1 - x}$.

20. $y = \log^4 (1 + x^2) \sqrt{1 - 2x}$.

* Some writers define $\log^2 x$ as $\log (\log x)$.

EXERCISE XI

Simple Exponential Functions. In the following formula v may be any function of x, the independent variable,

$$\frac{d}{dx} e^v = e^v \frac{dv}{dx}.$$

Differentiate the following:

1. $y = e^{x^2+2}$.
2. $y = e^{\log_3 x}$.
3. $y = e^{\sqrt{x}}$.
4. $y = e^{\sqrt{x^2+4}}$.
5. $y = e^{\log(x^2+x)}$.
6. $y = e^{\sqrt{1-x^2}}$.
7. $y = e^{\sqrt{3-2x}}$.
8. $y = e^{2/x-1}$.
9. $y = e^{3-1/x}$.
10. $y = e^{(1-t)/(1+t)}$.
11. $y = e^{1/(t^2+4)}$.
12. $y = e^{t \log t}$.
13. $y = e^{t^2/(2t+3)}$.
14. $y = e^{t\sqrt{1+t}}$.
15. $y = e^{\log_4(x^2+4)}$.
16. $y = e^{(x+1)/\sqrt{x}}$.
17. $y = e^{1/\sqrt{1-x}}$.
18. $y = e^{\sqrt{1-x}/x}$.
19. $y = e^{1/(3x^2-5x)}$.
20. $y = e^{t^2 \log 2t}$.
21. $y = e^{x^2 \log 1/x}$.
22. $y = e^{x+1/\sqrt{x}}$.
23. $y = e^{e^{x^2}}$.
24. $y = e^{\sqrt[3]{x^2+2x+3}}$.
25. $y = e^{\log^2 x}$.
26. $y = e^{(2x+1)/x}$.
27. $y = e^{(t\sqrt{1-t})/(1+t)}$.
28. $y = e^{(x^2+1)/(x\sqrt{1+x})}$.
29. $y = e^{1/\sqrt{9-x^2}}$.
30. $y = e^{\sqrt{(1-t^2)/(1+t^2)}}$.

EXERCISE XII

Exponential Functions. Base Constant. When the base of the exponential function is a constant, a, other than e, the derivative of the preceding exercise is simply multiplied by the $\log_e a$; that is,

$$da^v/dx = a^v \log_e a \, dv/dx.$$

Differentiate:

1. $y = a^{x^2}$.
2. $y = 2^{(t^2+2t)}$.
3. $y = 5^{t \log t}$.
4. $y = 2^{\sqrt{x}}$.
5. $y = 7^{-1/x}$.
6. $y = 5^{(t^2+3t)^2}$.
7. $y = 7^{\frac{1}{2}\log x^2}$.
8. $y = 5^{\sqrt{1-2x}}$.
9. $y = 4^{2x/(x^2-5)}$.
10. $y = 2^{(x+1)/(2x+3)}$.
11. $f(x) = 8^{1/(1-x)}$.
12. $f(t) = b^{\sqrt[3]{t^3-3t}}$.
13. $f(x) = a^{1/\sqrt{a^2-x^2}}$.
14. $y = 3^{\log(\log x)}$.
15. $y = 2^{1/\log t}$.
16. $y = c^{\log^2 t/2}$.
17. $y = a^{\frac{1}{2}\sqrt{x^2+5}}$.
18. $y = a^{3/\sqrt{x^2+5}}$.
19. $y = b^{(e^{1/x}+1)}$.
20. $y = (a+b)^{\sqrt{1-t^2}}$.
21. $y = a^{t^2} b^{t+1}$.
22. $y = 3^{\sqrt{x}} 2^{1/x}$.
23. $f(x) = 2^{\sqrt{x}+1/\sqrt{x}}$.
24. $f(t) = a^{t+1} b^{1/(t+1)}$.

EXERCISE XIII

Logarithmic Differentiation. It is desirable to take the logarithm of both sides to the base e before differentiating in the following cases: (a) where there is a "variable to a variable power" involved, and (b) in the case of products and quotients involving three or more terms. In every case be sure that the right-hand member has been written in the best form for differentiation before proceeding. (See Exercise VIII.) Apply this process in differentiating the following:

1. $y = \dfrac{x^2\sqrt{x^2+7}}{\sqrt{x^2+1}}$.

2. $y = \dfrac{1+x^2}{x^3\sqrt{1-x^2}}$.

3. $y = x^3(x^2+4)^{\frac{1}{2}}(x^3+3x)^{\frac{1}{3}}$.

4. $y = \dfrac{(2x-5)^3}{(1-5x^2)\sqrt{x^2-1}}$.

5. $y = \dfrac{3x^2\sqrt{2x+7}}{\sqrt{x^2+5}}$.

6. $y = (x+1)x^x$.

7. $y = (x+1)x^{1/x}$.

8. $y = x^5 \cdot 3^{x^2}$.

9. $y = x^{1/x}\log x$.

10. $y = \dfrac{1+4t^2}{t^2\sqrt{(1+t^2)^3}}$.

11. $x = t^2(1-t^2)^{\frac{1}{2}}(1-t^3)^{\frac{1}{3}}$.

12. $f(x) = \dfrac{2x^2}{(x-1)^2\sqrt{2x+4}}$.

13. $f(x) = 4x^3(1-2x)^7(x+5)^4$.

14. $x = \dfrac{2t^2(1-3t)^3}{(1-t+t^2)^2}$.

15. $x = \dfrac{2y^3\sqrt{6y+1}}{\sqrt{3y^2+4y+4}}$.

16. $x = 2t^2\sqrt[3]{3t+6}\,\sqrt{t^2+2t}$.

17. $y = \dfrac{x^2}{5\sqrt{6x+5}\cdot\sqrt[3]{3x^2+4}}$.

18. $y = \dfrac{2x\sqrt{3x^2+2x+1}}{\sqrt[3]{x^3-3x}}$.

19. $y = \dfrac{(3t^2-2)\sqrt{t^2+3}}{2t^2-5t+6}$.

20. $y = 7x\sqrt{3-4x}\,\sqrt{4x^3-3x^2}$.

21. $y = 4(1-x^2)\sqrt{1-x}\,\sqrt[3]{1+x}$.

22. $y = \sqrt[3]{\dfrac{(x-2)(x-3)}{(x-4)(x-5)}}$.

23. $y = \dfrac{(2-3t)^3\sqrt{1+t}}{(3-2t)^2\sqrt[3]{1-t}}$.

24. $y = \dfrac{x^3\sqrt{1+\log x}}{(1+x)^4}$.

EXERCISE XIV

General Exponential Functions. These functions are all variables to a variable power, or of the type u^v. They are to be differentiated by logarithmic differentiation. The student is cautioned to note the difference between v^n, a variable to a constant power; a^v, a constant to a variable power; and the present case, u^v, a variable to a variable power. In the last case only need we resort to logarithmic differentiation. Differentiate:

1. $y = x^{1/x}$.
2. $y = (1 - x)^{x^2}$.
3. $y = (1 + x)^{x^2}$.
4. $y = 2x^{\sqrt{8x}}$.
5. $x = 3(t^2 - 6t)^t$.
6. $y = 3x^{x^2+8}$.
7. $y = (1/x)^{\sqrt{x}}$.
8. $y = 3(x + 1)^{x+1}$.
9. $y = (x^2 - 4)^{1/2x}$.
10. $y = 2x^{x^3-8x}$.
11. $y = 4(3x^2 - 4x)^{3x}$.
12. $y = (1 + \log x)^{x^2}$.
13. $y = (2t + 5)^{t^2+8}$.
14. $y = x^{\sqrt{1-x}}$.
15. $y = t^{e^{t^2}}$.
16. $y = (1 - 3x)^{\log x}$.
17. $y = x^{x/(1+x)}$.
18. $y = 3t^{t^2+8t+1}$.
19. $y = (x \log x)^x$.
20. $y = (x + 1)^{e^x}$.
21. $y = x^{e^x+1}$.
22. $y = (\sqrt{x^2 + 4})^{x/2}$.
23. $y = x^2(x^2 + 4)^{2x}$.
24. $y = \sqrt[x]{3x^2 - 6x + 7}$.
25. $y = (2x + 7)^{1/(2x+1)}$.
26. $y = (1 - x^2)^{\sqrt{1-x}}$.
27. $y = [(1 + x)/(1 - x)]^{e^x}$.
28. $y = x^{\sqrt{(1+x)/(1-x)}}$.
29. $y = 2(x^2 + 4x + 3)^{x+1}$.
30. $s = [\log(2t + 1)]^t$.

EXERCISE XV

Simple Trigonometric Functions. The derivatives of the six trigonometric functions are

$$\frac{d}{dx}\sin v = \cos v\,\frac{dv}{dx}; \qquad \frac{d}{dx}\cot v = -\csc^2 v\,\frac{dv}{dx};$$

$$\frac{d}{dx}\cos v = -\sin v\,\frac{dv}{dx}; \qquad \frac{d}{dx}\sec v = \sec v \tan v\,\frac{dv}{dx};$$

$$\frac{d}{dx}\tan v = \sec^2 v\,\frac{dv}{dx}; \qquad \frac{d}{dx}\csc v = -\csc v \cot v\,\frac{dv}{dx}.$$

As a mnemonic fact it should be noted that the derivatives of the co-named functions have the *minus* sign. Also notice that the argument, that is, the angle, does *not* change in the process of differentiation. Do not forget the dv/dx. Differentiate the following:

1. $y = \sin x^3$.
2. $y = \sin(2x + 5)$.
3. $y = \sin(2/x)$.
4. $y = \sin\sqrt{x}$.
5. $y = \sin e^{x^2}$.
6. $y = \cos(3 - 2x)$.
7. $y = \cos(e^{2t})$.
8. $y = \cos\sqrt{1 - t}$.
9. $y = \sin(x^2 + 1/x)$.

10. $y = \cos(\log\sqrt{x})$.
11. $y = \tan(2x^2 + 3)$.
12. $y = \sec e^x$.
13. $y = \cot(1/x^2)$.
14. $y = \csc(x^2 + 1)$.
15. $y = \tan e^{1/t}$.
16. $y = \cos(e^{1/x} + e^{-1/x})$.
17. $y = \sin[(2x + 7)/(1 - 2x)]$.
18. $y = \cos(\log\sqrt{x^2 - 2x})$.
19. $y = \cos(\sin x^2)$.
20. $y = \cos(2e^x + 2e^{-x})$.
21. $y = 2\cos[(1-2x^2)/(3x^2-2)]$.
22. $y = \sec[1/(x^2 + 2)]$.
23. $y = \cos(t^2 - 1/t)$.
24. $y = \cot(7 - 2\theta^2)$.
25. $y = \cos 1/\sqrt{x}$.
26. $y = \csc[(1 - x)/(1 + x)]$.
27. $y = \tan 1/\sqrt{1 - x^2}$.
28. $y = \sec a^{x^2}$.
29. $y = \csc(1/x + 1)^2$.
30. $y = \tan\log[(1 - 3x)/(2x - 4)]$.
31. $y = \csc[(3x - 2)/(x^3 - 3)]$.
32. $s = \sin\log(2t + \sqrt{4t^2 - 1})$.
33. $x = \cot e^{1/(t+1)}$.
34. $\theta = \sec\log(t^3 + 3)$.
35. $\theta = \cos(2/x^2 + 7)$.

EXERCISE XVI

Powers of Trigonometric Functions. No new formulas are required, but care must be observed to distinguish between $\sin^3 x$ and $\sin x^3$, the former being of the type v^n, while the second is simply $\sin v$. Notice that the "argument" (the angle) *never* changes during the process of differentiation, and the differentiation of the argument is the final step. That is, in differentiating $\sin^3 2/x$, $2/x$ is the argument, and the differentiation of this will be the last step. Thus $\frac{d}{dx}\left(\sin^3\frac{2}{x}\right) = 3\sin^2\frac{2}{x}\cdot\cos\frac{2}{x}\cdot -\frac{2}{x^2} = -\frac{6}{x^2}\sin^2\frac{2}{x}\cos\frac{2}{x}$. Differentiate:

1. $y = \sin^3 x^2$.
2. $y = \cos^2\sqrt{x}$.
3. $y = \tan^4(1/x)$.
4. $y = \sec^5\log x$.
5. $y = \cot^3 e^{2t}$.
6. $y = 2\sin^4 e^{-3t}$.
7. $s = 3\tan^2(x + 3)$.
8. $\theta = 2\cos^3(1/3x)$.
9. $y = 4\sqrt{\sin 2x}$.
10. $y = 6\sqrt[3]{\cos x^3}$.
11. $y = \csc^2(3/x)$.
12. $y = \cot^2(1 - 2x)$.
13. $y = \sin^2[(x - 4)/(x + 2)]$.
14. $y = \sqrt{\sin 3x^2}$.
15. $y = \tan^3(e^{x/2} - e^{-x/2})$.
16. $y = 2\sec^2(\log 1/x)$.
17. $y = 3\cos^4(x\log x)$.
18. $y = \cot^3 a^{x^2}$.
19. $y = \csc^3\sqrt{1 - 2x^2}$.
20. $y = \sin^4[x/(1 - x)]$.
21. $x = 2\sec^2[1/(t^2 - 1)]$.
22. $f(t) = \sin^4(t^2 - 1/t)$.
23. $s = \cot^2[(2t^2 - 5)/(3 - t)]$.
24. $x = \sin^2(ye^y)$.

25. $y = \sin^2(\sin 2x + 1)$.

26. $f(s) = \csc^4(3 - 2/s)^2$.

27. $f(x) = \cot^2 e^{1/(x^2-1)}$.

28. $f(x) = 4\tan^3\sqrt{e^{2x} + e^{-2x}}$.

29. $y = \tan^3(\log \sin x^2)$.

30. $y = \sin^3 x^x$.

31. $y = \cos^4(t+1)^t$.

32. $y = \csc^2 x^{1-2x}$.

33. $y = \cot^3(te^{1/t})$.

34. $y = \sec^2[3\,e^{1/(t+2)}]$.

35. $y = 1/\tan^2\sqrt{1-t^2}$.

36. $y = 4/\sin(t^2 \log t)$.

EXERCISE XVII

Forms Involving Trigonometric Functions. In the following products, quotients, exponentials, etc., no additional formulas are required.

1. $y = \sin 3t \cos 2t$.

2. $y = e^{2t}\sin t/2$.

3. $y = e^{\sin 2/t}$.

4. $y = \sin t/2 \tan t/2$.

5. $y = x^{\sin x}$.

6. $y = 2\,a^{\sec x}$.

7. $y = e^{t/2}\tan t^2$.

8. $y = 2^{\tan 3x}$.

9. $y = e^{\sin 2\theta}\cos 2\theta$.

10. $x = \log \sin^3 2\theta$.

11. $s = \log \sec^2 t/2$.

12. $y = (\sin t)^t$.

13. $y = \log(\sec 3x + \tan 3x)$.

14. $s = (t/3)^{\sec 2t}$.

15. $y = \sin 2x \tan 1/x$.

16. $y = \sin^2 2/x \tan^2 x/2$.

17. $y = e^{\tan x}\cos x^2$.

18. $y = 7^{-x/4}\cos(3 - 2x)$.

19. $y = \log \tan 1/2x$.

20. $y = (\sin t)^{\tan t}$.

21. $f(t) = e^{\sec 2t}/\sin 2t$.

22. $f(t) = e^{\sin 2t}\log \sin 2t$.

23. $y = \log(\sin\theta \sin 2\theta \sin 3\theta)$.

24. $y = \log_{10}(\sin\theta\cos 3\theta)$.

25. $y = (\cos x)^{1+\cos x}$.

26. $y = \sqrt{x}\cos(1 - x^2)$.

27. $y = 2(\sin 3t)^{2t+1}$.

28. $f(x) = (\sec 1/x)^{2x}$.

29. $f(t) = e^{\tan 3t}\sin 3t$.

30. $f(t) = 3^{\sin t}t^{\log t}$.

EXERCISE XVIII

Simple Inverse Trigonometric Functions. The formulas for the derivatives of the inverse trigonometric functions are

$$\frac{d}{dx}\operatorname{arc}\sin v = \frac{dv}{dx}\Big/\sqrt{1-v^2}, \qquad \frac{d}{dx}\operatorname{arc}\tan v = \frac{dv}{dx}\Big/(1+v^2),$$

$$\frac{d}{dx}\operatorname{arc}\cos v = -\frac{dv}{dx}\Big/\sqrt{1-v^2}, \qquad \frac{d}{dx}\operatorname{arc}\cot v = -\frac{dv}{dx}\Big/(1+v^2),$$

$$\frac{d}{dx}\operatorname{arc}\sec v = \frac{dv}{dx}\Big/v\sqrt{v^2-1}, \qquad \frac{d}{dx}\operatorname{arc}\csc v = -\frac{dv}{dx}\Big/v\sqrt{v^2-1}.$$

The student should notice that the derivatives of the "co-named" functions differ from the others only by the *minus* sign. Differentiate:

1. $y = \text{arc} \sin 2x$.
2. $y = \text{arc} \cos x^2$.
3. $y = \text{arc} \tan 3x$.
4. $y = \text{arc} \sec x^3$.
5. $y = \text{arc} \sin 1/x$.
6. $y = \text{arc} \cot e^x$.
7. $y = \text{arc} \tan 2/x$.
8. $y = \text{arc} \csc 2x^2$.
9. $y = \text{arc} \cos 1/2x$.
10. $y = \text{arc} \sin \sqrt{x}$.
11. $y = \text{arc} \sec (1-x)$.
12. $y = \text{arc} \sin e^{x^2}$.
13. $y = \text{arc} \tan (\log x)$.
14. $y = \text{arc} \tan \frac{1}{2}(e^{2t} - e^{-2t})$.
15. $y = \text{arc} \csc \frac{1}{2}(e^{2/t} + e^{-2/t})$.
16. $y = \text{arc} \cos [(2x+3)/(3x+2)]$.
17. $y = \text{arc} \sin [(1-t)/(1+t)]$.
18. $f(t) = \text{arc} \tan [(t-1)/\sqrt{t}]$.
19. $f(x) = \text{arc} \cos (1/\sqrt{1+x})$.
20. $\theta = \text{arc} \tan \log \sqrt[3]{t}$.
21. $f(x) = \text{arc} \sec [(x^2+1)/2x]$.
22. $f(x) = \text{arc} \cot [(x^2-1)/2x]$.
23. $f(x) = \text{arc} \tan (\sin x)$.
24. $f(x) = \text{arc} \cos \log (x^2-1)$.
25. $y = \text{arc} \sec \frac{1}{2}(a^{x^2} + a^{-x^2})$.
26. $y = \text{arc} \sin \tan (2x+5)$.
27. $y = \text{arc} \sin [2/(e^{3x^2} - e^{-3x^2})]$.
28. $y = \text{arc} \sec 2^{\sqrt{t}}$.

EXERCISE XIX

Forms involving Inverse Trigonometric Functions. In the following no new formulas are involved. Simplify algebraic results as much as possible.

1. $y = x \text{ arc} \cos 2x$.
2. $y = e^x \text{ arc} \sin e^x$.
3. $y = \log \text{arc} \tan x/2$.
4. $y = 3 \log (t/\sqrt{t^2+4}) + \frac{5}{2} \text{ arc} \tan t/2$.
5. $y = x \text{ arc} \sin x/2 + \sqrt{4-x^2}$.
6. $y = 8x/(x^2+4) - 4 \text{ arc} \tan x/2 + x$.
7. $f(t) = \log [(t^2 - 6t + 13)/t^2] + 10 \text{ arc} \tan \frac{1}{2}(t-3)$.
8. $x = 2\sqrt{t-4} + 2 \text{ arc} \tan \frac{1}{2}\sqrt{t-4}$.
9. $y = \frac{1}{3}x^3 \text{ arc} \tan x - \frac{1}{6}x^2 + \frac{1}{6}\log(x^2+1)$.
10. $f(x) = \frac{9}{2} \text{ arc} \sin x/3 - \frac{1}{2}x\sqrt{9-x^2}$.
11. $f(x) = \sqrt{2-x-x^2} + 3 \text{ arc} \sin \sqrt{(x+2)/3}$.
12. $y = (1-x^2)^{\frac{3}{2}} \text{ arc} \sin x - x + x^3/3$.
13. $s = 2\sqrt{t-9} + 3 \text{ arc} \tan \frac{1}{3}\sqrt{t-9}$.
14. $y = \sqrt{t^2-1}/2t^2 - \frac{1}{2} \text{ arc} \sec t$.
15. $y = x \text{ arc} \sin (\cos 2x)$.
16. $y = e^{2x}(\text{arc} \sec 2/x)^3$.
17. $y = \sqrt{x^2-1} \text{ arc} \csc x$.
18. $y = (1-x^2)/x^3 \cdot \text{arc} \sin 1/x - (x^2-1)^{\frac{3}{2}}/3x^3$.

EXERCISE XX

Function of a Function. In many problems a certain variable y is a function of v, while v in turn is a known function of a third variable, x or t. Obviously one could first eliminate v, and express y as a function of x; then differentiate as before. But it is frequently much easier to make use of the theorem

$$\text{(A)} \qquad dy/dx = dy/dv \cdot dv/dx.$$

(The student is reminded that this theorem is *not* proved by merely canceling out the dv on the right-hand side.) Find dy/dx (or dy/dt) in each of the following, using theorem (A).

1. $y = v^{\frac{3}{2}} + 4$, $v = 1 + \log x$.
2. $y = (1 - t^2)^{\frac{3}{2}}$, $t = 1/x$.
3. $y = e^v$, $v = 3^{x^3}$
4. $y = \sin\sqrt{v}$, $v = \cos x^2$.
5. $y = z^2 + 2z$, $z = \arcsin(x/a)$
6. $y = 2z/(z + 2)$, $z = (2x + 5)/(x - 1)$.
7. $y = v^2/(v - 1)$, $v = e^{x+3}$.
8. $y = \log t^3$, $t = x/(x + 4)$.
9. $y = \log[(3v^2 + 5)/v]$, $v = e^{2x}$.
10. $y = z\log z$, $z = \sec 1/x$.
11. $y = \log\dfrac{(2 + z)^2}{1 - z}$, $z = \sin x$.
12. $y = 2v - 4\arctan v/2$, $v = \sqrt{e^x - 1}$.
13. $y = \log(1 + v^2) + \sqrt{1 + v^2}$, $v = \tan x$.
14. $y = \log\left(3v + \sqrt{9v^2 + 4}\right)$, $v = \frac{2}{3}\tan t$.
15. $y = \sqrt{v} + \arctan\sqrt{v}$, $v = \sqrt{1 + x^2}$.
16. $y = 2\arctan\sqrt{(1 - x)/(1 + x)}$, $x = \cos t$.
17. $y = \arctan\left(\sqrt{1 + v^2} - v\right)$, $v = \cot x$.
18. $y = \log[(3v + 2)/(2v + 3)]$, $v = \sec x + \tan x$.
19. $y = e^t + e^{2t}$, $v = \log(x - x^2)$.
20. $y = \sqrt{v} - \log\left(\sqrt{v} + 1\right)$, $v = e^{2x}$.

EXERCISE XXI

Implicit Functions. In an implicit function, the function is not explicitly expressed in terms of the independent variable, but the relation between x and y is given by an equation involving both variables. In differentiating products, quotients, etc., involving two variables in this way, the ordinary formulas for products, etc., are applicable, giving directly an equation involving x, y, and dy/dx, which can be solved for dy/dx. The result in general will involve both variables.

Illustration: $xy + x^2y + y^2 = 0.$

Differentiating:

$$x\,dy/dx + y + x^2dy/dx + 2\,xy + 2\,y\,dy/dx = 0.$$

Solving: $dy/dx = -(y + 2\,xy)/(x + x^2 + 2\,y)$. Find dy/dx in each of the following:

1. $xy + x^2 + y^2 = 0.$
2. $x^2y - 4\,y^2 = 0.$
3. $3\,x^2y + 4\,x - 2\,y = 0.$
4. $y^3 - 3\,xy + x^2 = 0.$
5. $x^2y + 3\,y + x^2 = 16.$
6. $x^3 + 2\,xy - y^2 = 4.$
7. $x^2 + 2\,xy - 3\,y^2 = 0.$
8. $x^{\frac{1}{2}} + y^{\frac{1}{2}} = 6.$
9. $x^{\frac{2}{3}} + y^{\frac{1}{2}}x^{\frac{1}{2}} = 0.$
10. $3\,x^3 - xy^2 - 2\,y = 0.$
11. $x^2y + 2\,xy + y^2 = 0.$
12. $y^2 - 2\,xy^2 + x^2y^2 + x^2 = 0.$
13. $xy^2 + 2\,x^2y + 3\,xy = 5.$
14. $y^3 + 3\,x^2y + xy^2 - 3\,x + 5 = 0.$
15. $e^{x+y} + e^{x-y} = 0.$
16. $\sin^2 x \cos^2 y = 0.$
17. $x^2 = y^2 \log xy.$
18. $e^{2x} - \log(y^2 + 4) + 7 = 0.$
19. $e^{2x+y} + \sin(x + 2\,y) = 0.$
20. $3^y 2^x = \text{arc tan } y^2.$
21. $\text{arc sin } xy - \text{arc tan} \log y = 0.$
22. $(\sin y)^{x^2} - (\cos x)^{y^2} = 0.$
23. $(\sin x)^y = y^{\sin x}.$
24. $\sqrt{x^2 + y^2} = \log \sqrt{x^2 - y^2}.$
25. $\log \sin xy - e^{x+y} = 0.$
26. $x^2 \sin y - y^2 \sin x = 16.$
27. $e^{2/x} \sin y/2 - e^{x/2} \sin 2/y = 0.$
28. $y/\sqrt{x^2 + y^2} + \text{arc tan } y/x = 0.$
29. $2 \log(x + y) + \log x/y = 0.$
30. $x^{\log y} + y^{\log x} = x + y.$

EXERCISE XXII

General Review of Formal Differentiation. Find the derivative in each of the following:

1. $y = 2^{x^2} + \sqrt{3\,x} + 1/2\,x^2 + e^{\sqrt{x}}.$
2. $f(x) = 2\,x/(4 + x^2) + \text{arc tan } x/2.$
3. $f(x) = \frac{1}{2}\,x\sqrt{x^2 - 1} + \frac{1}{2}\log\left(x + \sqrt{x^2 - 1}\right).$
4. $x = (2\,t^2 - 1)/3\,t^3 \cdot \sqrt{t^2 + 1}.$
5. $y = \log\left(\sqrt{x} + \sqrt{x + 4}\right).$
6. $y = 3\,x^3 + 1/3\,x^3 + \sqrt[3]{3\,x} + 3^{x^3} + e^{x^3}.$
7. $y = x^2 - 1/2\,x^2 + \sqrt{2\,x} + 4^{x/2} + e^{2/x}.$

8. $x^2y + xy^2 - 4x + 4y - 5 = 0$.

9. $y = (1 + e^{x/2})/(1 - e^{x/2})$.

10. $y = \frac{1}{2}(x-4)\sqrt{8x - x^2} + 8 \operatorname{arc\,sin} \frac{1}{4}(x-4)$.

11. $y = x \log(4x^2 - 1) - 2x - \frac{1}{2}\log[(2x-1)/(2x+1)]$.

12. $y = x\log(4x^2+1) - 2x + \operatorname{arc\,tan} 2x$.

13. $y = (x+1)\operatorname{arc\,tan}\sqrt{x} - \sqrt{x} + 4$.

14. $y = 2 \operatorname{arc\,sin} t/2 - \frac{1}{2} t\sqrt{4 - t^2}$.

15. $x = \sqrt{t^2 - 4}/8t^2 + \frac{1}{16}\operatorname{arc\,sec} t/2$.

16. $y = \sqrt{1-x}\operatorname{arc\,sin}\sqrt{x} - \sqrt{x}$.

17. $y = \log\cot(1 - x^2)$.

18. $y = \sec x \tan x + \log(\sec x + \tan x)$.

19. $y = (x^3 + 2x)\sqrt{x^2+4} - 8\log\left(x + \sqrt{x^2+4}\right)$.

20. $y = \log_6\left[\sqrt[3]{3x^2+7}/(x^2\sqrt{2x+5})\right]$.

21. $y = (\sin 1/x + \cos 1/x)/e^{1/x}$.

22. $y = (\sin x^2)^{2x-1}$.

23. $y = \operatorname{arc\,tan} x^x$.

24. $y = \operatorname{arc\,sin}[4/(e^{x^2} + 4e^{-x^2})]$.

25. $y = \log[(e^{2x} + e^{-2x})/(1 + e^x)]$.

26. $y = (1 - e^{-8x})/(1 + e^{8x})$.

27. $y = (2x^3 + 5x)\sqrt{x^2+1} + 3\log\left(x + \sqrt{x^2+1}\right)$.

28. $y = \log_x(x+2)$.

29. $y = \operatorname{arc\,cot}\frac{1}{2}(e^{1/x^2} - e^{-1/x^2})$.

30. $y = [(1-x)/x]^{(1-x)/x}$.

31. $y = \log(x+1)^3(x-1)(x^2+1) + 5\operatorname{arc\,tan} x$.

32. $y = \frac{1}{2}(1+x^2)\log(1+x^2) - x^2/2$.

33. $y = \left(648 - 108\sqrt{x} - 18x - 5x\sqrt{x}\right)/6\sqrt{2 - \sqrt{x}}$.

34. $y = \log_4\left(\sqrt{2x^2-3}/\sqrt[3]{1-x^3}\right)$.

35. $y = \operatorname{arc\,sin} 2/x + \operatorname{arc\,cot} x/2$.

36. $y = [\log(e^x + 1)]/(e^x + 1)$.

37. $y = e^{-t/3}(\sin 3t - \cos 3t)$.

38. $y = \log(e^{2x} + a^{8x})^2$.

39. $y = e^{x^2}\sec^2 2x$.

40. $y = 1/x(\log 1/x - 1)$.

41. $y = (1+2x)\log(1+2x) - 2x$.

42. $y = (e^{2x} + 3)^x$.

43. $y = \cos^3 3x \tan 3/x$.

44. $y = \sin^8 3/x \sec x/3$.

45. $y = e^{-t/4}\tan^2 4t$.

46. $y = e^{-t/4}(2\sin 2t - 3\cos 2t)$.

47. $x\sin y + y\cos x + y^2 - xy = 0$.

48. $xy - x\sec y - y\tan x - 16 = 0$.

49. $e^{xy} + \log xy + \sin xy - x^2 = 0$.

50. $\sin(x + 2y) + \cos(2x + y) - 16 = 0$.

51. If a, b, and c are the roots of $f(x) = 0$, prove that

$$f'(x)/f(x) = 1/(x-a) + 1/(x-b) + 1/(x-c).$$

CHAPTER III

SIMPLE APPLICATIONS. SUCCESSIVE DIFFERENTIATION

EXERCISE XXIII

Direction of Curves. In the following draw the curve and the tangents at the points considered.

1. Find the slope of the curve $y = \sin x$ for the points where $x = 0$, $\pi/6$, $\pi/4$, $\pi/2$, π.

2. Find the points at which the curve $x^2 + y^2 - 6x - 4y + 12 = 0$ is (a) parallel to the x-axis; (b) parallel to the y-axis.

3. At what points on the circle $x^2 + y^2 = 25$ is a particle moving parallel to the straight line $3x - 4y + 7 = 0$?

4. At what points on the curve $y = x^3 - 12x + 4$ is the slope of the tangent equal to 15?

5. Find the points at which the curve $6y = 2x^3 - 15x^2 + 42x - 30$ has (a) inclination $= 45°$; (b) slope $= 3$.

6. Find the points where the tangent to the curve $y = \frac{1}{3}x^3 - x^2 + \frac{1}{3}$ (a) has an inclination of $3\pi/4$, and (b) is parallel to the x-axis.

7. Two particles always remaining in the same vertical line follow paths given by the curves $y = 2x^2 - 8x + 1$ and $2y = x^2 + 8x - 5$. When will the two particles be moving parallel?

8. Find the coördinates of the points on the curve $xy = 16$ at which the normals pass through the origin.

9. What is the direction of the curve $y = x^3 - 6x^2 + 11x - 6$ at the points where it croses the x-axis?

10. Find the points at which the curve $6y = 2x^3 - 3x^2 - 42x + 60$ has (a) inclination $= 135°$; (b) slope $= 5$.

11. At what points on the curve $y = x/(1 - x^2)$ has the tangent an inclination of $45°$?

12. At what points is the tangent to the curve $y = x^3 + 1$ (a) parallel to the line $3x - y = 1$? (b) perpendicular to the line $x + 12y - 5 = 0$?

13. Where is the tangent to $y = 6(x - 6)/(x^2 - 2x - 8)$ parallel to the x-axis?

14. Find the points at which the curve $2y = 2x^3 - x^2 + 1$ is (a) parallel to the line $10x - y - 5 = 0$; (b) perpendicular to the line $x + 2y + 3 = 0$.

15. Find the points at which the circle $x^2 + y^2 + 4x - 6y = 3$ is (a) parallel to the x-axis; (b) parallel to the y-axis.

16. Where is the tangent to the ellipse $4y^2 - 4xy + 2x^2 - 2x = 8$ parallel to the line $x - 2y + 4 = 0$?

17. Where on the ellipse $y^2 - 2xy + 2x^2 - 2y - 7 = 0$ has the tangent an inclination of 45°?

18. Find the points for which the tangent to the curve $y = \sin x + \cos x$ is (a) parallel to the x-axis; (b) has an inclination of 45°.

19. Given the curve $y = \frac{2}{3}x^3 - \frac{1}{2}x^2 + 1$. Find (a) the inclination at the points where $x = 0, 1, 2$; (b) the points for which the slope is 2.

20. At what points on the curve $x^4 - 8xy^2 + y^4 = 0$ is the tangent parallel to the y-axis?

21. Two points always remaining in the same vertical line follow paths given by the curves $y = x^2 + x - 6$ and $2y = x^2 - 2x - 8$ respectively. When will the points be moving (a) in parallel directions? (b) in perpendicular directions?

22. What is the inclination of the curve $y = 4x/(x^2 + 4)$ at the origin? at the point (2, 1)?

23. Find the points at which the curve $y = 2\sin x + 3\cos 2x$ has slope = 0.

24. Find the points at which the curve $y = 5\sin^2 x + 2\cos 4x$ has slope = 0.

25. Find the coördinates of a point on the curve $y = 2\sin^2 x$ where the inclination is $\pi/3$.

26. At what points of the curve $y = \sin 3x - 3\sin x$ is the slope = 0?

27. Find the points for which the curve $y = 2\sin^2 x + 3\cos 2x$ has slope = 0.

28. At what point on the curve $x^3 + xy^2 + 4y^2 - 4x^2 = 0$ is the tangent parallel to the y-axis?

29. Find the angles at which the curve $y^3 = x^2 - 7x$ cuts the line $y = 2$.

30. At what angles does the curve $4y = x^3 - 4x$ cut the x-axis?

31. Where on the cissoid $y^2 = x^3/(4 - x)$ is the slope of the tangent equal to 2?

32. Where on the curve $y^2 = x^2(4 - x)$ is the tangent parallel to $x + 2y + 5 = 0$?

33. At what point on the curve $xy^2 = 16$ does the normal go through the origin?

34. Given the equation of the family of curves $4y = 4ce^{-2x} + 2x^2 - 2x + 1$, where c is the variable parameter. Find the locus of points on the curves of this family at which the tangents are parallel to the x-axis.

35. Given the curve family $2y = 2ke^x + \sin x + \cos x - 2$, where k is the parameter. Find the locus of points on the curves of this family where the tangents have an inclination of 45°.

EXERCISE XXIV

Algebraic Curves. Tangents, Normals, and Angles of Intersection. Since dy/dx is equal geometrically to the slope of the tangent to the algebraic curve represented by $y = f(x)$[or by $F(x, y) = 0$], we can at once apply our knowledge of calculus to the following problems concerning algebraic curves. Evaluating the derivative for the values (x_1, y_1) of any point gives us the slope m of the tangent line. Then $y - y_1 = m(x - x_1)$ is the equation of this tangent line, and $y - y_1 = -1/m \cdot (x - x_1)$ is the equation of the normal to the curve at the point (x_1, y_1).

Now if two curves are given, we can find the value of dy/dx, or m, for each curve at this point. Call them m_1 and m_2, and then $\tan \theta = (m_1 - m_2)/(1 + m_1 m_2)$ will give us the angle θ between the two curves at their point of intersection. In the following examples find (a) the equation of the tangent and normal to each curve at one point of intersection, (b) the angle between the two curves. Draw figures. If the origin is a point of intersection, take another point. Calculate angles in degrees.

1. $\begin{cases} 2y = x^2, \\ 4y = x^2 + 4. \end{cases}$ **2.** $\begin{cases} 4y = x^2 + 4, \\ x^2 = 8 - 2y. \end{cases}$ **3.** $\begin{cases} 2y = x^2, \\ x^2 - 4x + 2y = 0. \end{cases}$

4. $\begin{cases} x^2 - 2x = 2y - 1. \\ x^2 - 6x + 2y + 5 = 0. \end{cases}$ **6.** $\begin{cases} y = x^2 - 3x, \\ 9y = 2x^2 - 4x + 20. \end{cases}$

5. $\begin{cases} y = x^2 - 2x - 3, \\ y = 3x - x^2. \end{cases}$ **7.** $\begin{cases} 4y = 2x^2 - 3x, \\ 4y = x^2 + 4. \end{cases}$

8. $\begin{cases} x^2 = 4y, \\ 2y^2 = x. \end{cases}$ 10. $\begin{cases} 2y^2 = x^3, \\ y^2 = 32x. \end{cases}$ 12. $\begin{cases} y^2 = 8 - 4x^3, \\ y^2 = 4x^3. \end{cases}$

9. $\begin{cases} x^2 = 8y^3, \\ 4y = 9x - x^2. \end{cases}$ 11. $\begin{cases} y^2 = 16x, \\ y = x^2 - 2x. \end{cases}$ 13. $\begin{cases} 8y = x^3, \\ 4y = 4x - x^2. \end{cases}$

14. $\begin{cases} y^2 = x^3 - 4, \\ x^2 + y^2 - 6x + 4 = 0. \end{cases}$ 15. $\begin{cases} x^2 + y^2 + 2x - 24 = 0, \\ 4x^2 + y^2 + 8x - 48 = 0. \end{cases}$

16. $\begin{cases} x^2 = 16y, \\ y^2 - 6y = x. \end{cases}$ 17. $\begin{cases} xy = 12, \\ x^2 + y^2 = 25. \end{cases}$ 18. $\begin{cases} y^2 = x^3, \\ 7x - 3y - 4 = 0. \end{cases}$

19. $\begin{cases} 10x^2 + y^2 - 34 = 0, \\ 3y - 2x^2 = 0. \end{cases}$ 21. $\begin{cases} 2x^3 - y^2 + 4y - 4 = 0, \\ y^2 - 4y - 8x + 4 = 0. \end{cases}$

20. $\begin{cases} y^2 - 4y - x = 0, \\ x + y^2 - 6 = 0. \end{cases}$ 22. $\begin{cases} y^2 = x^3/(4 - x), \\ y^2 = 4x - 4. \end{cases}$

For the remaining curves calculate the length of the subtangent, s_t, and of the subnormal, s_n, as well as the equations of the tangent and normal at the points indicated. The student should not try to remember any formulas for lengths, as they are readily calculated from the figure. The observing pupil will soon note that $s_t = y_1/m$ and $s_n = y_1 m$, where y_1 is the given ordinate.

23. $xy^2 = 16$, at $(2, 2\sqrt{2})$. Show that the normal passes through the origin.

24. $y^2 - 2x^2 - 3xy - 4x = 0$, at the points where $x = 2$.

25. $x^2 + y^2 - 6x - 16 = 0$, at the point where it crosses the positive end of the y-axis.

26. $y^2 = x^3/(4 - x)$, at $(2, 2)$.

27. $x^2 + 2y^2 - 2xy - x = 0$, at $(1, 0)$.

28. $y^3 = x^2 - 6x$, at $x = 2$.

29. $y = (x - 3)^2$, at the points where the absolute length of the subtangent is 2 units.

30. Given the three curves $y = 4x^2$, $y = 2x^3$, and $y = x^4$. Show that for all values of x the lengths of the subtangents form a harmonic progression.

EXERCISE XXV

Tangents and Normals. Parametric Equations. When x and y are both given as functions of a parameter t (or θ),

$$\frac{dy}{dx} = \frac{dy}{dt} \div \frac{dx}{dt}.$$

Then the slope at a point determined by any value of t, say $t = a$, is found by evaluating dy/dx for $t = a$. This is m, and we continue as in the preceding exercise.

In each of the following find the equation of the tangent and of the normal for the indicated values of the parameter, and also calculate the length of the subtangent and of the subnormal.

1. $x = t^2 - 1$, $y = t^3 - 4$, at $t = 2$.

2. $x = t^2 + 1$, $y = t + 3$, at $t = 2$.

3. $x = 3t^2 - 1$, $y = 1 - 2t$, at $t = 1$.

4. $x = 3 - 2t$, $y = t^3 - 3$, at $t = 1$.

5. $x = t^2 + 2$, $y = t^3 + 1$, at $t = 1$.

6. $x = 4t + t^2$, $y = 2t^2$, at $t = 1$.

7. $x = 2t$, $y = 3/t$, at $t = 3$.

8. $x = \sin\theta$, $y = \cos 2\theta$, at $\theta = \pi/6$.

9. $x = 4\sec\theta$, $y = 2\tan\theta$, at $\theta = \pi/6$.

10. $x = \tan t$, $y = \sin 2t$, at $t = 0, \pi/6, \pi/4$.

11. $x = \sec\theta$, $y = \tan\theta$, at $\theta = \pi/4$.

12. $x = t^3 + 2t^2 + 3$, $y = t^3 + 3t + 5$, at $t = -2$.

13. $x = 1/(t + 1)$, $y = \sqrt{5 - t^2}$, at $t = 0, 1, 2$.

14. $x = 1/(1 - 2t)$, $y = \sqrt{2t^2 + 1}$, at $t = 0, 2$.

15. $x = 1/t + 1$, $y = t^2 + 1$, at $t = 1$.

16. $x = 3t/(1 + t^3)$, $y = 3t^2/(1 + t^3)$, at $t = 2$.

17. $x = \sin^3\theta$, $y = \cos^3\theta$, at $\theta = \pi/3, \pi/4$.

18. $x = 4\cos\theta$, $y = 2\sin^2\theta$, at $\theta = \pi/3, \pi/4$.

EXERCISE XXVI

Polar Curves. The student should avoid the common error of saying that the slope of the tangent in polar coördinates is $d\rho/d\theta$, as this is not true. The two important angles in polar curves are (a) the angle at a point P on the curve, between the radius vector and the tangent to the curve at P, which we will call ψ; and (b) the angle τ between the tangent line at a point and the polar axis. These angles are given by the formulas (as proved in the text)

$$\tan\psi = \rho \div d\rho/d\theta = k, \quad \text{and} \quad \tau = \psi + \theta,$$

where θ is the coördinate angle.

Obviously if two intersecting curves are given, the angle of intersection, ϕ, is $\psi_1 - \psi_2$, where ψ_1 and ψ_2 are the values of ψ for the two curves at the point of intersection; or

$$\tan\phi = (k_1 - k_2)/(1 + k_1 k_2).$$

Show that the following curves intersect at right angles:

1. $\rho = a(1 + \sin\theta)$, $\rho = a(1 - \sin\theta)$.
2. $\rho = 2\sin\theta$, $\rho = 2\cos\theta$.
3. $\rho = 4\sec^2\theta/2$, $\rho\sin^2\theta/2 = 8$.

Find the angle of intersection between the following pairs of curves:

4. $\rho = 4\cos\theta$, $\rho = 4(1 - \cos\theta)$.
5. $\rho = 2(1 + \cos\theta)$, $\rho = 8\cos\theta$.
6. $\rho = 4(1 - \sin\theta)$, $\rho = 4\sin\theta$.
7. $\rho = \sin\theta$, $\rho = \sin 2\theta$.
8. $\rho = \sin\theta$, $\rho = \cos 2\theta$.
9. $\rho = \cos 2\theta + 1$, $\rho = \sin 2\theta$.
10. $\rho^2 \sin 2\theta = 8$, $\rho = 2\sec\theta$.
11. $\rho^2 \sin 2\theta = 4$, $\rho^2 = 16\sin 2\theta$.
12. $\rho = 4\sin\theta$, $\rho = 4\sin\theta\tan\theta$.
13. $\rho = \sin\theta + \cos\theta$, $\rho = 2\sin\theta$.

It is sometimes desirable to know the polar coördinates of the extreme points on a curve. As is clear from a figure, at the extreme high and low points $\tau = 0$, or $\tan\tau = 0$; at the extreme right and left points $\tau = \pi/2$. In each of the following examples calculate the extreme points:

14. $\rho = 2\sin^3\theta/3$. Also find τ, the inclination of the tangent for $\theta = \pi/4$.

15. $\rho = \sin\theta + \cos\theta$.
16. $\rho = 4(1 - \cos\theta)$.
17. $\rho = \sin 2\theta$.

18. Given the cardioid $\rho = 5(1 + \cos\theta)$ and the point P where $\tan\theta = \frac{3}{4}$. Find the angle which the tangent to the curve at P makes with the polar axis. Sketch your figure and mark angles clearly.

EXERCISE XXVII

Velocity and Acceleration. Rectilinear Motion. When the distance s of a particle from an arbitrarily fixed origin is given as a function of the time t, the velocity v, at any time t, is given by the value of the derivative ds/dt at that time. The acceleration is dv/dt, obtained by differentiating a second time. It is clear then that the particle comes to rest and usually changes direction whenever $ds/dt = 0$, and that its speed is either maximum or minimum whenever $dv/dt = 0$. In the first

12 problems calculate the position, velocity, and acceleration for the given value of t; calculate also when the particle first comes to rest.

1. $s=2t^3-15t^2+36t+10, t=1.$

2. $s=-2t^3+15t^2+36t, t=2.$

3. $s=24t+3t^2-t^3, t=3.$

4. $s=2t^3-27t^2+120t, t=3.$

5. $s=120t-3t^2-2t^3, t=2.$

6. $s=60t+3t^2-2t^3, t=3.$

7. $s=t^3+1/t, t=2.$

8. $s=t\sin^2 t, t=1, t=\pi/4.$

9. $s=e^{t^3-1}, t=1.$

10. $s=\log(1+t^2), t=3.$

11. $s=5^{t^2}, t=2.$

12. $s=\sin t+\cos 2t, t=1.$

In the remaining problems calculate by the aid of a table (a) the position, velocity, and acceleration at the given time t, and (b) the two smallest values of t for which the velocity will be zero.

13. $x=e^{-t/10}\sin 2t, t=5.$

14. $x=5e^{-t}\cos t, t=1.$

15. $x=2e^{t/2}\sin 2t, t=1.$

16. $x=4e^{-t/3}\sin t, t=2.$

17. $x=4e^{-t/4}\sin 2t, t=1.$

18. $x=2e^{-t/5}\sin t, t=4.$

19. $x=4e^{-t/2}\cos 3t, t=1.$

20. $x=5e^{-t/4}\cos \pi t/2, t=2.$

21. $x=6e^{-t/10}\cos \pi t/4, t=5.$

22. $x=2e^{-t/4}\sin \pi/2\,(t+1), t=2.$

23. $x=10e^{-t/2}\cos \pi/3\,(t+2), t=3.$

24. $x=6e^{-t/5}\sin \pi/4\,(t-1), t=3.$

25. Assuming the formula $s=v_0t-16t^2$, calculate the time of flight of a particle projected vertically upward with a velocity of 288 ft./sec. What is the velocity at the end of the time?

26. Two particles are moving in different media, projected however from a common level. If the equations of motion are $s=v_0t-16t^2$ and $s=v_0t-12t^2$, and each is projected vertically upward with initial velocity $v_0=96$ ft./sec., which will go the higher and how much? When and with what velocity will each return to the level from which projected?

27. The equation of motion for a falling body may be written $s=v_0t+16t^2$. If one particle is dropped from rest and 1 second later a second is projected directly downward with $v_0=64$ ft./sec., will it overtake the first? If so, when and with what velocity will it pass the first? If started two seconds later than the first, show that they will remain a constant distance apart.

28. A man in a balloon drops a bomb from a height of 2400 ft. and at the same time a shot is fired directly upward with $v_0=400$ ft./sec. How far from the earth will the shot meet the bomb, assuming the aim to be true?

EXERCISE XXVIII

Multiple Roots of Equations. If the equation $f(x)=0$ has a as a double root, then $x-a$ is a common factor of both $f(x)$ and $f'(x)$. Hence the roots of the H.C.F of $f(x)$ and $f'(x)$ are *multiple* roots of the original equation, and if the H.C.F is simply a constant, the equation has no multiple roots. Use this knowledge to calculate the roots of the first 16 equations.

1. $x^4-2x^3-7x^2+20x-12=0.$
2. $4x^4-20x^3+37x^2-30x+9=0.$
3. $8x^3+28x^2+30x+9=0.$
4. $18x^3-3x^2-4x+1=0.$
5. $8x^4-4x^3-6x^2+5x-1=0.$
6. $4x^4-4x^3-15x^2+16x-4=0.$
7. $2x^4-x^3-9x^2+13x-5=0.$
8. $9x^4+6x^3-11x^2-4x+4=0.$
9. $4x^4-99x^2-245x-150=0.$
10. $x^4+4x^3+2x^2-4x-3=0.$
11. $9x^4-6x^3-8x^2+6x-1=0.$
12. $x^4+6x^3+5x^2-24x-36=0.$
13. $4x^4+12x^3+x^2-12x+4=0.$
14. $x^4-4x^3-6x^2+36x-27=0.$
15. $4x^4-27x^2-25x-6=0.$
16. $x^5+5x^4-5x^3-25x^2+40x-16=0.$

17. In the following equations determine without solving which have multiple roots and which have not:

(a) $x^4-5x^3+2x^2+20x-24=0.$
(b) $x^3-6x^2+11x-6=0.$
(c) $x^3-x^2-x-2=0.$
(d) $4x^4-19x^2-3x+18=0.$
(e) $x^4-19x^2+6x+72=0.$
(f) $81x^4-135x^3+54x^2-12x-8=0.$
(g) $x^4+2x^2-x+2=0.$
(h) $x^4+4x^2-4x+16=0.$
(i) $8x^4-28x^3+18x^2+27x-27=0.$
(j) $8x^5-4x^4-10x^3-x^2+4x-1=0.$
(k) $4x^4-8x^3+11x^2-16x+6=0.$
(l) $36x^4+36x^3-31x^2-4x+3=0.$

18. Use the method of this exercise to find the condition that $ax^2+bx+c=0$ shall have equal roots.

19. What is the condition on a and b that the equation $x^3+x^2+ax+b=0$ shall have a double root?

20. What is the condition that $x^3+ax^2+x+b=0$ shall have a double root?

EXERCISE XXIX

Successive Differentiation. These exercises embody no new principles, as the idea of a second derivative was used in acceleration. The process may be repeated indefinitely, although in all polynomials the $(n+1)$th and all succeeding derivatives will be zero, n being the degree of the polynomial. The student should note in each case whether there is any law governing the successive derivatives which would enable him to write them down without actually differentiating repeatedly. Find the second and third derivatives in all cases. Also the nth derivative for the first six.

1. $y = \sin x + \cos x$.
2. $y = 1/x$.
3. $y = e^{2x}$.
4. $y = \cos 2x$.
5. $y = 5/x^2$.
6. $y = e^{3x+1}$.
7. $y = x^4 - 12x^3 + 48x^2 - 50$.
8. $y = x^3 - 2\sin x$.
9. $y = x \arctan x$.
10. $y = e^{-t}\sin t$.
11. $y = 1/\sqrt[3]{x^2}$.
12. $y = \sqrt[3]{x^2}$
13. $y = e^{4+t^2}$.
14. $y = x^3e^x$.

15. $y = \log[(2+3x)/(2-3x)]$.
16. $y = xe^{-x^2}$.
17. $y = (\log x)/x$.
18. $y = x^2\log x$.
19. $y = \arccos x$.
20. $y = t\sin t + \cos t$.
21. $y = 4e^{-t/2}\sin 2t$.
22. $y = 6e^{t/2}(\sin 2t + \cos 2t)$.
23. $x = 3e^{-t/3}\cos t/2$.
24. $y = 8e^{-t/4}(\sin t/2 + 2\cos t/2)$.
25. $y = \frac{1}{3}\tan^3 x - \tan x + x$.
26. $y = \log[(1+\sin x)/(1-\sin x)]$.
27. $y = x \arcsin x$.
28. $y = \arcsin(1-x^2)$.
29. $y = \log\left(x+\sqrt{1+x^2}\right)$.
30. $y = 1/x \cdot \sin 2/x$.

EXERCISE XXX

Successive Differentiation. Implicit Functions. In finding the second and higher derivatives when the function is an implicit one, it is generally desirable to find dy/dx first and then differentiate both sides with respect to x. Of course the right-hand member in this second differentiation will contain dy/dx, the value of which was found in the first operation. A second method is to perform the second differentiation implicitly before solving for dy/dx explicitly. Find d^2y/dx^2 unless otherwise stated.

1. $x^3 - xy + y^3 = 0.$
2. $y^2 - 3xy - 2 = 0.$
3. $x^2 + 2y^2 - 2xy - x = 0.$
4. $x^3 + y^3 - 3x = 0.$
5. $x^2 - 2xy + 4y = 0.$
6. $y^2 + 2x^2 - 4y + 2x - 2 = 0.$
7. $x^4 + 2x^2y^2 - 48 = 0.$
8. $\sin x + \cos y = 1.$
9. $e^x + e^y = 4.$
10. $x + \cos(x + y) = 0.$
11. $x + \sin(x + y) = 0.$
12. $e^{xy} - x = 0.$
13. $e^{3x-2y} - 4x = 0.$
14. $(y - \sin x)^2 = 4x.$
15. $\sqrt{x + y} = e^x.$
16. $e^{\sin x}\cos y = 0.$
17. $e^x + e^y = e^{x+y}.$
18. $x^2 \log y = 4.$
19. $y^2 \log x = 4.$
20. $ye^x = x + y.$

Find d^3y/dx^3 for the following: Nos. 2, 5, 12, and 17.

EXERCISE XXXI

Successive Differentiation of Parametric Forms. We have already shown that $dy/dx = dy/dt \div dx/dt$. To find d^2y/dx^2 there are two common methods. (1) Use the somewhat cumbersome formula

$$\frac{d^2y}{dx^2} = \left(\frac{dx}{dt}\cdot\frac{d^2y}{dt^2} - \frac{dy}{dt}\cdot\frac{d^2x}{dt^2}\right) \Big/ \left(\frac{dx}{dt}\right)^3.$$

(2) If $dy/dx = F(t)$,

$$d^2y/dx^2 = F'(t)\,dt/dx,$$

where $F'(t)$ is the derivative of $F(t)$ with respect to t. *Do not forget the dt/dx.* The second method is more easily remembered. Find d^2y/dx^2 for each of the following:

1. $x = 2t + 1,\ y = t^3/3.$
2. $x = t^2 - 1,\ y = t^3 - 2t.$
3. $x = 3t,\ y = 3/t.$
4. $x = 3t^2 + 2t,\ y = t^3.$
5. $x = \log t,\ y = 1/t.$
6. $x = \sin 2t,\ y = \sin^2 t.$
7. $x = a\cos t,\ y = b\sin t.$
8. $x = \cos 2t,\ y = \sin^2 t.$
9. $x = e^{-at},\ y = e^{at}.$
10. $x = \log t,\ y = 1/(1 - t).$
11. $x = e^t,\ y = \sin t.$
12. $x = \log(t + 2),\ y = 1/(t + 2)^2.$
13. $x = \sec^2 t,\ y = \tan t.$
14. $x = t - \sin t,\ y = 1 - \cos t.$
15. $x = \frac{2}{3}\sqrt{2t^3},\ y = \frac{1}{2}t^2.$
16. $x = 3/(3 - t),\ y = 2/(2 - t).$
17. $x = \sin t - t\cos t,\ y = \cos t + t\sin t.$
18. $x = 2\sin t/(1 + 2\cos t),\ y = \cos t/(1 + 2\cos t).$
19. $x = \cos t/(1 + 2\cos t),\ y = 2\sin t/(1 + 2\cos t).$
20. The folium $x = 3t/(t^3 + 1),\ y = 3t^2/(t^3 + 1).$

CHAPTER IV

MAXIMA AND MINIMA

EXERCISE XXXII

Maximum and Minimum Points of Algebraic Curves. Calculus furnishes two methods of determining relative maximum and minimum points of curves. In any case, both at a maximum and at a minimum point the first derivative is zero, or infinite. Hence the values of the independent variable x which make the first derivative vanish or become infinite will be called *critical values.* Then to select the maximum and minimum values we can

(1) Substitute in dy/dx a value of x a little smaller and then one a little larger than the critical value being considered. If the value of dy/dx changes from $+$ to $-$, this critical value gives a maximum point; if it changes from $-$ to $+$, a minimum point. (If the sign does not change, it gives a point of inflection with horizontal tangent.)

(2) If x is a critical value which makes $dy/dx = 0$, we can proceed as follows: Take the second derivative and evaluate for the critical value of x. If this value is negative, the corresponding point is a maximum; if it is positive, the corresponding point is a minimum. If it is zero, it is usually a point of inflection.

At every point of inflection d^2y/dx^2 equals zero, and this knowledge enables us to find the points of inflection. In the above discussion of maxima and minima the word "curve" may be replaced by "function," and "point" by "value of the function."

Find the coördinates of the maximum, minimum, and inflection points of the following curves. Use this knowledge to sketch the graphs.

1. $y = 2x^3 - 9x^2 - 24x - 12.$
2. $y = 2x^3 - 3x^2 - 36x + 25.$
3. $3y = x^3 - 6x^2 - 36x + 30.$
4. $y = x^3 + 12x^2 + 36x - 50.$
5. $y = 3x^3 - 9x^2 - 27x + 30.$
6. $y = 15 + 9x - 3x^2 - x^3.$
7. $3y = x^3 - 3x^2 - 9x + 3.$
8. $2y = 2 + 3x - 4x^2 - x^3.$
9. $y = x^4 - 8x^3 + 22x^2 - 24x + 12.$
10. $6y = x^4 - 16x^3 + 88x^2 - 192x + 192.$
11. $5y = 2 + 24x^2 - x^4.$
12. $y = x^4 - 4x^3 - 2x^2 + 12x + 6.$
13. $y = x^5 - 7x^4 + 19x^3 - 25x^2 + 16x - 4.$
14. $y = x^5 - 5x^4 + 5x^3 + 1.$
15. $8y = 4x^5 + 5x^4 - 60x^3 + 110x^2 - 80x + 1.$
16. $y = x^6 - 6x^4 + 4x^3 + 9x^2 - 12x + 4.$
17. $y = 6x/(x^2 + 1).$
18. $y = x + 4/x.$
19. $y = (x-2)^2(x-3)^2.$
20. $y = (x-4)^2(x-3)^2.$
21. $y = x^2 + 128/x.$
22. $y = (x^2 + 9)/x.$
23. $y = x - 108/x^2.$
24. $y = 6x/(x^2+1).$
25. $y = x^3 + 48/x.$
26. $y = (x^3 - 16)/x.$
27. $xy = x^2 + x + 1.$
28. $xy + x^2 + 2x + 4 = 0.$
29. $xy - x^2 + y = 0.$
30. $y = x(x-2)^2(x+4)^2.$
31. $y = (2x+1)^3(x+2)^2.$
32. $y = (x^2 - x + 9)/(x^2 + x + 9).$
33. $y = (x^2 - 3x - 19)/(x + 4).$
34. $y = (x^2 - 3x + 2)/(x^2 + 3x + 2).$
35. $y = 8x/(x^2 + 2x + 4).$
36. $y = 4(2 - x)/(x^2 + 4).$
37. $y = x/(x^2 + 6x + 9).$
38. $y = (x^2 + x - 1)/(x^2 - x + 1).$
39. $y = (x + 1)/(x^2 - 4x + 1).$
40. $y = x^3/(2x + 4)$, the trident of Newton.

EXERCISE XXXIII

Maxima and Minima. Transcendental Functions. In finding the relative maximum and minimum points on a transcendental curve, it is generally the simpler method to determine the nature of the points given by the critical values, by means of the sign of the second derivative. If both the first and second derivatives vanish, the higher derivatives determine the nature of the point. Determine the coördinates of the maximum and minimum points on the following curves. Solve graphically or by the aid of tables where necessary.

1. $y = \sin^3 x$.
2. $y = e^x \cos x$.
3. $y = 5 \text{ arc tan } x - x$.
4. $y = \sin^2 x + \sin 2x$.
5. $y = \frac{1}{4} \cos^2 x + \sin^2 x$.
6. $y = e^x \cos 2x$.
7. $y = \sin x + \cos 2x$.
8. $y = \sin 2t + \cos t$.
9. $y = 2 \sin t + \sin 2t$.
10. $y = 2 \sin t + \cos 2t$.
11. $y = \cos t - \frac{1}{2} \cos 2t$.
12. $y = 2 \cos t - \sin 2t$.
13. $x = e^{-t} \sin 2t$.
14. $y = 2e^{-t/4} \sin 2t$.
15. $x = 8e^{t/2} \sin t$.
16. $y = 5e^{-t/2} \cos 2t$.
17. $x = 2e^{-t/8} \cos t/2$.
18. $x = 5e^{-t} \sin (t + \pi/3)$.
19. $x = 4e^{-t/8} \sin (t + 1)$.
20. $x = 4e^{-t/2} \cos 3t$.
21. $x = 4e^{2t} (\sin t + \cos t)$.
22. $x = 5e^{t/2} (\sin 2t + 2 \cos 2t)$.
23. $x = 4e^{-t/5} \sin \pi/2 \cdot (t - 1)$.
24. $x = 6e^{-t/10} \cos \pi/2 \cdot (t + 1)$.

EXERCISE XXXIV

Maxima and Minima. Problems. In all practical problems involving maxima and minima, it is necessary first of all to note carefully the function or quantity, F, which is to be a maximum or minimum. This quantity must then be expressed in terms of some one independent variable, say x. Then the maximum or minimum values of the function will be for those values of x which make $dF/dx = 0$. The nature of the problem usually enables one to tell in advance whether the result is a maximum or minimum, without applying any formal test. The following examples give rise to algebraic functions only.

1. A rectangular flower bed is to contain 432 sq. ft. It is surrounded by a walk which is 4 ft. wide along the sides and 3 ft. wide across the ends. If the total area of bed and walk together is a minimum, what are the dimensions of the flower bed?

2. A sheet of paper for a poster contains 12 sq. ft. The margins at top and bottom are 4 in. and on the sides 3 in. What are the dimensions if the printed area is a maximum?

3. A letter Y of total height 16 in. and width across the top of 12 in. is to be made. Find the length of the vertical stem AB, if the length of the stem plus that of the two equal branches is a minimum.

4. Divide the number 8 into two such parts that (a) the sum of the squares may be a minimum, and (b) the product of the two parts may be a maximum.

5. A rectangular box is to be made from a sheet of tin 15×24 in. by cutting a square from each corner and turning up the sides. Find the side of the squares cut out if the volume of the box is a maximum.

6. A rectangular garden is to be laid out along a neighbor's lot and is to contain 432 sq. rd. If the neighbor pays for half the dividing fence, what should be the dimensions of the garden so that the cost to the owner of inclosing it may be a minimum?

7. A box with a square base and lid is to contain 360 cu. ft. If the bottom costs 4¢, the lid 6¢, and the sides 3¢ per square foot, what are the dimensions for minimum cost?

8. A right prism, the base of which is an equilateral triangle, has a volume of 2 cu. ft. Find the edge of the base for minimum *total* surface.

9. A vertical circular cylindrical water tank is to be built to contain $3000\,\pi$ cu. ft. If the cost of the bottom, including foundations, is \$4 per square foot, that of the sides \$2 per square foot, and it is roofed with a hemispherical dome costing \$1 per square foot, find the dimensions for minimum cost.

10. The cross section of an open irrigation canal is to be of the form of an isosceles trapezoid with the slope of the sides equal to $\frac{4}{3}$. The area of the cross section is to be 448 sq. ft. What are the dimensions if the retaining surface (bottom plus sloping sides) is to be a minimum?

11. A farmer estimates that if he digs his potatoes now he will have 120 bu. worth \$1 per bushel, but if he waits, the crop will grow 20 bu. per week, while the price will drop 10¢ per bushel per week. When should he dig them to get the largest cash returns?

12. A telephone company agrees to put in a new exchange for 100 subscribers or less at a uniform charge of \$20 each. To encourage a larger list of subscribers they agree to deduct 10¢ from this uniform charge for each subscriber in excess of the 100 (i.e. if 110 subscribe, the flat rate would be \$19). What number of subscribers would give the telephone company the maximum gross income?

13. A spring is located 40 rd. at right angles from a point B on a straight road. A traveler at A, 100 rd. down the road from B, wishes to reach the spring in the shortest time. If he can walk 5 mi. per hour on the road and 4 mi. per hour across the fields, where should he leave the road?

14. The perimeter of a sector of a circle is 16 ft. What is the radius of the circle if the area of the sector is a maximum? ($A = \frac{1}{2} R \times \text{arc.}$)

15. The shore of a lake is a straight line, and two towns A and B are located 4 and 8 mi., respectively, from D and E, the nearest points on the lake shore, DE being 9 mi. One pumping station on the lake shore is to supply both towns with water. Where must it be located so that the length of the mains to the towns may be a minimum?

16. A farmer is 12 mi. from A, the nearest point on a straight railway. The railroad company agrees to put in a siding anywhere he designates, and to haul his produce to B, 80 mi. along the track from A, for 5¢ per ton per mile. If he can haul by wagon for 13¢ per ton per mile, where should the siding be located that the cost of transportation of his crops to B may be a minimum? (Assume a straight road from farm to new siding.)

17. The area of a sector of a circle is 16 sq. ft. What is the radius of the circle if the perimeter of the sector is a minimum? ($A = \frac{1}{2} R \times \text{arc.}$)

18. A telephone company finds there is a net profit of $15 per instrument if an exchange has 1000 subscribers or less. If there are over 1000 subscribers, the profits per instrument decrease 1¢ for each subscriber above that number. How many subscribers would give the maximum net profit?

19. A window composed of a rectangle surmounted by an equilateral triangle is 15 ft. in perimeter. Find its dimensions if it admits the maximum amount of light. Answer to two decimal places.

20. A window is in the form of a rectangle surmounted by an isosceles triangle, the altitude of which is $\frac{3}{8}$ of its base. If the perimeter is 30 ft., find the dimensions for admitting the maximum amount of light.

21. A model is to be constructed composed of a rectangular parallelopiped with square base topped by a pyramid of altitude $\frac{3}{8}$ of the side of the base. Its total volume is to be 3 cu. ft. All the surface except the base is to be covered with gold leaf. Find its dimensions for minimum cost of gold leaf.

22. A model of the same character as in example 21 is to be constructed with a volume of 7 cu. ft. Find its dimensions for minimum total surface.

23. Given a sphere of radius 6 in. Calculate the quantities desired for the inscribed solids specified:

(a) Altitude of right circular cylinder of maximum volume.
(b) Altitude of right circular cylinder of maximum total surface.
(c) Altitude of right cone of maximum volume.
(d) Altitude of circumscribed cone of minimum volume.

24. A printing company agrees to print 50,000 posters (or less) at the rate of \$10 per 1000. If the number needed exceeds 50,000, they agree to deduct 10¢ per 1000 on the whole contract, for each 1000 in excess of 50,000. What number of posters would give the printing company the maximum cash receipts, and what average price would they receive?

25. If the radius of a cylinder is 2 ft. and increasing 1 ft. per hour while the altitude is 4 ft. and decreasing 1 ft. per hour, when will the volume be a maximum? How much will it have increased?

26. A grain dealer has 10,000 bu. of new wheat worth \$1 per bushel in an elevator. There is a shrinkage of 200 bu. per month, a fixed storage and interest charge of \$120 per month, and a promised advance of 5¢ per bushel per month. If these conditions prevail indefinitely, how long should he hold the wheat?

27. A ship sailing due north 12 mi. per hour sights another ship dead ahead at a distance of 10 mi. and sailing east at the rate of 9 mi. per hour. If each ship keeps its course, what will be the least distance between them at any time, and how soon will this position be reached?

28. Two boats are at A and B, A being 65 mi. due north of B. The first boat sails southeast $10\sqrt{2}$ mi. per hour, and the other due east 25 mi. per hour. When will they be nearest each other, and how far apart will they be?

29. The sum of the perimeters of a circle and of a square is constant. Show that for minimum total area the diameter of the circle equals the side of the square.

30. The sum of the areas of a sphere and of a cube is a constant. What is the relation between the diameter of the sphere and the edge of the cube if the total volume is a minimum?

31. A right circular cylindrical tank is to hold 5000 cu. ft. What are its dimensions for minimum cost if the bottom costs \$1 per square foot, and the sides \$1.50 per square foot? (Two decimals.)

32. The perimeter of a triangle is 16 in. If one side is 6 in., what are the other two for maximum area?

33. A triangle ABC has base $AB = 6$ in., and altitude 8 in. Take D any point in AB. Draw a line parallel to AB cutting the other sides in E and F respectively. What is the altitude of DEF if its area is a maximum?

34. A given isosceles triangle has base 20 ft. and altitude 8 ft. What are the dimensions of the maximum inscribed parallelogram, one side coinciding with the base of the triangle, if the acute angle of the parallelogram is arc tan $\frac{4}{3}$?

35. Given OX and OY, two perpendicular lines (which may be regarded as coördinate axes), and a point $P(a, b)$. Draw a line through P cutting the positive ends of the axes at A and B. Calculate the intercepts of this line on OX and OY

(a) when the area OAB is a minimum;
(b) when the length AB is a minimum;
(c) when the sum of the intercepts is a minimum;
(d) when the perpendicular distance from O to AB is a maximum.

Show that these results are all the same if $a = b$.

36. Given the parabola $y^2 = 8x$, and the point $P(6, 0)$ on the axis. Find the coördinates of the points on the parabola nearest to P.

37. On a line 6 in. long construct a triangle of altitude 4 in. such that the perimeter may be a minimum.

38. Given a triangle ABC, with D and E on sides BA and BC respectively. If $BA = a$, $BC = b$, and DE is the shortest line dividing the triangle into two equivalent parts, calculate BE and BD.

39. A steel girder 32 ft. in length is to be moved on rollers along a passageway and into a corridor 4 ft. in width at right angles with the passageway. How wide must the passageway be in order that the girder may go around the corner? In the solution the horizontal width of the girder may be neglected.

40. In using the formula $S = P/W$ to find the specific gravity of a substance, the following pairs of values for P and W were obtained from different specimens: (27, 5), (31.2, 6), (35.7, 7), and (42.4, 8). If the best value for S is the one which makes $(P_1 - SW_1)^2 + (P_2 - SW_2)^2 + (P_3 - SW_3)^2 + (P_4 - SW_4)^2$ a minimum, calculate the best value and see how much it differs from the arithmetical mean of the four values of S obtained from the separate experiments.

41. Assume Hooke's law that the tension in an elastic cord or spring is proportional to the amount it has been stretched beyond its normal length; that is, $T = kx$. In an experiment with a spring, 6 lb. stretched it 2.1 in.; 10 lb., 3.4 in.; and 12 lb., 4.4 in. If the best value of k is the one which makes the sum of the squares of the differences $T - kx$ a minimum, calculate the best value of k and compare it with the arithmetical mean of the values given by the individual measurements. (See preceding example.)

42. Assume that under similar conditions a gasoline launch uses a constant amount of gasoline per mile, or that $y = kx$, where y is the number of miles and x the amount of gasoline in gallons. The observed values on certain trips under similar conditions were 6 gal. in 20 mi., 8.5 gal. in 27 mi., 3.5 gal. in 12 mi., and 11 gal. in 35 mi. What is the best value for k, using the same definition as in Problem 41? How much does it differ from the arithmetical mean?

43. Assume the law $PV = k$. A series of experimental values of P and V are (2, 4), (3, 2.8), and (4, 1.9). If the best value of k is the one which makes the sum of the squares of the differences $(P_1 - k/V_1)^2 + (P_2 - k/V_2)^2 + (P_3 - k/V_3)^2$ a minimum, calculate k. Compare it with the arithmetical mean of the three separate values.

EXERCISE XXXV

Maxima and Minima. Problems (*continued*). In the following practical problems the function which is to be a maximum or minimum may be either algebraic or transcendental. The student should note that our methods require that the quantity which is to be a maximum or minimum in these problems should be expressed as a function of one independent variable or parameter. A wise choice of parameter is frequently of the utmost importance.

1. The turning effect of a ship's rudder is shown theoretically to be $k \cos\theta \sin^2\theta$, where θ is the angle the rudder makes with the keel, and k is a constant. For what value of θ is the rudder most effective?

2. Given a triangle ABC, with points D and E on BA and BC respectively. BA is 8 units long and BC is 10. If DE divides the triangle ABC into 2 equivalent parts, find the distance BD and BE if DE is a minimum.

3. In a triangle the base a and the angle A opposite are fixed. Find the other two sides (a) when the area is a maximum; (b) when the perimeter is a maximum. (Use x, a variable angle, as parameter.)

4. Given the ellipse $x^2 + 3y^2 = 28$, and the two points (5, 1) and (– 4, 2) on it. Find the coördinates of a third point on the ellipse so that the triangle having these three points as vertices may be a maximum.

5. A regular right hexagonal prism has a volume of 36 cu. in. What are its dimensions if the total surface is a minimum?

6. $ABCD$ is a rectangle, and a straight line APQ cuts BC in P and DC extended in Q. Locate the point P so that the sum of the areas of the two triangles ABP and CPQ may be a minimum.

7. Two lines AB and AC of fixed length meet in A, forming an angle of fixed magnitude. Draw a line through A (dividing the angle BAC) such that the sum of the projections of AB and AC on this line may be a minimum.

8. Find the dimensions of the cylinder of maximum volume which can be inscribed in a sphere of radius 6 in. (Use the angle θ subtended by the radius of the base of the inscribed cylinder as a parameter. Then $r = 6 \sin \theta$, $h = 12 \cos \theta$.)

9. Solve example 8 if the convex surface of the cylinder is to be a maximum, using the same parameter.

10. A hexagonal tower is surmounted by a pyramidal roof with a 30° inclination. Find the relative dimensions

(a) for minimum surface when the volume is given;
(b) for maximum volume when the surface is given.

11. An angle a is given and remains constant. On one leg of the angle two points A and B are taken, CA being 4 in. and CB being 9 in. How far from C on the other leg must a point D be taken so that the angle ADB is a maximum?

12. A tank is formed by a vertical cylinder with a conical lid. The altitude of the lid is three fourths the radius of the base. The volume of the cylindrical portion is 104π cu. ft. The material for bottom and sides costs $2 per square foot and that for the lid $1 per square foot. What are the most economical dimensions for the tank?

13. A cylinder is surmounted by a cone the altitude of which is equal to its base diameter. Find the relative dimensions of the solid

(a) for maximum volume when the surface is given;
(b) for minimum surface when the volume is given.

14. A rectangular parallelopiped with a square base is surmounted by a square pyramid, the altitude of which equals $\frac{3}{8}$ of a side of its base. Find the relative dimensions

(a) for maximum volume when the total surface is given;
(b) for maximum volume when the surface not including base $= 36$ sq. ft.;
(c) for minimum total surface when the volume is given.

15. Two right triangles ABC and $A'B'C'$ are such that legs AB and $A'B'$ are 3 and 4 in. respectively, legs $BC + B'C' = 9\frac{1}{3}$ in. What are the lengths of BC and $B'C'$ if the sum of the two hypotenuses is a minimum?

16. The volume of a right cone is given. What is the relation between its altitude and radius when (a) the lateral surface is a minimum? (b) the total surface is a minimum?

17. Two points A and B are 100 ft. apart and there are street lights at these points of relative intensity 8 : 27. Assuming the law that the amount of light at a point is inversely proportional to the square of the distance from the source of light, how far from A on the path AB will the total amount of light from the two sources be a minimum? (See example 60.)

18. A produce house has 10,000 lb. of beef worth 16¢ in cold storage. If the beef loses 100 lb. per week in weight, the fixed charges are $60 per week, and the price advances 1¢ per pound per week, how long should the meat be held, and what is the selling price?

19. There are 20,000 bu. of wheat worth 85¢ per bushel in a certain elevator. If the shrinkage from all causes is 80 bu. per week, the fixed costs are $100 per week, and the price is advancing 1¢ per bushel per week, how long should the wheat be held, and what is the selling price per bushel? (Do not count interest on the money.)

20. The distance between the centers of two spheres of radius a and b respectively is c. Find from what point P on the line of centers AB the greatest amount of spherical surface is visible. (The superficial area of a zone of height h is $2\pi ah$, where a is the radius of the sphere.)

21. The cost of fuel for running a train is proportional to the square of the speed generated in miles per hour and costs $16 per hour at 16 mi. per hour. What is the most economical speed, fixed charges $100 per hour?

22. If the cost of running a steamboat is proportional to the cube of the speed generated, what is the most economical speed to run it against a 4 mi. per hour current?

23. If the cost of fuel for running a steamboat is proportional to the square of the speed generated, and is $4 per hour when going 6 mi. per hour, and the fixed charges are $16 per hour, what is the most economical speed against a $3\frac{1}{2}$ mi. per hour current?

24. What is the most economical speed when the boat in example 23 is going with the current? Compare the net progress per hour in the two cases.

25. A railway is to be built from A to B. A straight line L divides the plain region in which A is located from the hill region in which B is located. If A and B are each 12 mi. from D and E, the nearest points on L, and the distance DE is 25 mi. (along L), locate a point Q where the tracks should cross L, assuming the relative cost per mile on the plain and through the hills to be in the ratio 3 : 4. What is the length of the track and the change in direction at Q?

26. Given a sphere of radius 6 in., calculate the following (assume inscribed solids unless otherwise stated):

(a) Dimensions of maximum parallelopiped with square base.

(b) If one edge of the parallelopiped is 4 in., what are the other two for maximum volume?

(c) Dimensions of parallelopiped with square base and maximum surface.

(d) Altitude of maximum right cone.

(e) Altitude of right cone of maximum convex surface.

(f) Altitude of right cone of maximum entire surface.

(g) Altitude of maximum right cone with vertex at center of sphere.

(h) Altitude of right cone with vertex at center and maximum total surface.

(i) Altitude of circumscribed right cone of minimum convex surface.

(j) Altitude of circumscribed right cone of minimum total surface.

(k) Altitude and edge of base of right pyramid with square base and maximum volume.

(l) Altitude of square-based pyramid of maximum lateral surface.

(m) Altitude of square-based pyramid of maximum total surface.

(n) Altitude of maximum tetrahedon with equilateral base.

(o) Altitude of circumscribed pyramid with square base and minimum volume.

(p) Altitude of circumscribed tetrahedron with equilateral base and minimum volume.

27. A wall, with its top 8 ft. above the level of the water, surrounds an ice pond, and the ice house stands back 27 ft. horizontally from the wall. A regular incline of minimum length runs from the water to an opening in the side of the ice house, just resting on the top of the wall. What is the length of the incline, and how high is the opening above the level of the water? (Use the inclination as a parameter.)

28. A beam is to be carried into a mine where it is necessary to turn from a corridor of uniform width 8 ft. into one of width 6 ft. at right angles to it. If the beam is kept horizontal, what is its maximum length, disregarding thickness?

29. A mural painting 24 ft. high has its lower edge 8 ft. above the level of the observer's eye. How far away from the wall should he stand so as to get the most favorable view of the entire picture?

30. In a park there is a circle 40 ft. in radius, surrounded by a flat border of flowers, and an electric light is immediately above the center. How high above the ground should this light be placed so that there may be a maximum illumination of the inner flowers? Assume the ordinary laws that the intensity of light is proportional to the sine of the angle of incidence and inversely proportional to the square of the distance from the source of light.

31. Given a constant angle a, with its vertex at A. A fixed point M lies within the angle a, and a line drawn through this point cuts the sides of the angle at B and C respectively. Show that the area of the triangle is a minimum when M bisects BC.

32. A circular sector has a given perimeter. Prove that its area is a maximum when the angle of the sector is 2 radians, and that the maximum area equals the square of the radius.

33. The range, height, and time of flight of a projectile are given by

$$R = (v^2 \sin 2\,\theta)/g,\ \ h = (v^2 \sin^2 \theta)/2\,g,\ \text{ and } t = (2\,v \sin \theta)/g,$$

where v is a constant, the initial velocity, and g is the gravitational constant. Find the angle of projection θ which will make each of these in turn a maximum.

34. If a projectile is fired from O so as to strike an inclined plane which makes a constant angle α with the horizontal at O, the range is given by the formula $R = [2\,v^2 \cos\theta \sin(\theta - \alpha)]/g \cos^2\alpha$, where v and g are constants as in example 33 and θ is the angle of elevation. Calculate the value of θ, giving the maximum range up the plane.

35. When a load is being pulled up an inclined plane of constant inclination α, by a force making an angle θ with the horizontal, the efficiency of this device as a machine is given by the formula

$$E = [\cos(\alpha + \beta - \theta)\sin\alpha]/[\sin(\alpha + \beta)\cos(\alpha - \beta)],$$

where β, the angle of friction, is also a constant. At what angle must the force be applied for maximum efficiency?

36. For a square-threaded screw with pitch θ and angle of friction ϕ the efficiency is given by the formula $E = \tan\theta/[\tan(\theta + \phi) + f]$, where f is a constant. Find the value of θ for maximum efficiency when ϕ is a known constant angle.

37. A given weight W is to be raised by the aid of a lever weighing n pounds per linear foot. The fulcrum is at one end of the lever, and the weight W is suspended at a fixed distance a from that end. Find the length of the lever so that the power required to raise the weight may be a minimum.

38. The formula for the efficiency of a screw in mechanics is

$$E = h(1 - h\mu)/(h + \mu),$$

where μ is the constant coefficient of friction and h is the tangent of the angle of pitch of the screw. For what value of h is the efficiency a maximum?

39. The velocity of a wave of length λ in deep water is given by the formula $V = \sqrt{\lambda/a + a/\lambda}$, where a is a known constant. What is the length of the wave for minimum velocity?

40. Assuming that the formula for y, the total waste due to heat, depreciation, etc., which occurs in an electric conductor, is $y = C^2R + 17^2/R$, where R is the resistance in ohms per mile, and C the current in amperes, find the relation between C and R for minimum waste if the current is kept constant. (17 is an arbitrary value for k.)

41. Experiments on the explosion of mixtures at the pressure of one atmosphere lead to the empirical formula $p = 83 - 3.2x$, where p is the highest pressure produced in the explosion and x is the volume of air, together with products of previous explosions, added to 1 cu. ft. of coal gas before explosion. Now if the work done by a gas engine is roughly proportional to px, what value of x will make this a maximum?

42. The velocity equations for chemical reactions in which the normal course is disturbed by autocatalysis are, for reactions of the first order, $dx/dt = kx(a - x)$ and $dx/dt = k(b + x)(a - x)$, where k, a, and b are constants. Find in each case the values of x which give maximum velocities.

43. An electric current flows about a coil of radius r and exerts a force F on a small magnet the axis of which is on a line drawn through the center of the coil and perpendicular to its plane. This force $F = x/(r^2 + x^2)^{\frac{5}{2}}$, where x is the distance to the magnet from the center of the coil. Show that F is a maximum for $x = r/2$.

44. If VC denotes the input of a continuous current dynamo, δ the constant losses due to friction, iron, etc., C the current, V the voltage, and τC^2 the variable losses, the efficiency formula is

$$E = 1 - \delta/(VC) - (\tau C^2)/(VC).$$

Show that if the voltage is constant, the efficiency is a maximum when $\delta = \tau C^2$.

45. The power P transmitted by a belt is given by the formula $P = V(T - WV^2/g)$, where V is the linear velocity of the belt in meters per second, W is the weight per linear meter of the belt in kilograms, $g = 9.81$ being in meters per second², and T is the original tension in the belt while at rest. For what value of V is P a maximum?

46. Given an inclined plane of variable inclination α and angle of friction ϕ. If a load is to be elevated by means of this plane and a constant horizontal force, find α for maximum efficiency E, assuming the formula $E = (\tan\alpha)/\tan(\alpha + \phi)$.

47. Given the law for the deflection α in a tangent galvanometer, namely, $\tan\alpha$ is proportional to the current, or $\tan\alpha_1/\tan\alpha_2 = C_1/C_2$. Show that $\alpha_1 - \alpha_2$ is a maximum when α_1 and α_2 are complementary.

48. The intensity J of an alternating current is given by the formula $J = J_0 \sin(kt - \phi)$, where J_0 is the maximum value of J and ϕ is the phase angle. In a coil with a coefficient of self-induction L, the induced counter-electromotive force equals $L\,dJ/dt$. At what times is this induced E.M.F. a maximum? (An alternative form is $J = J_0 \sin(2\pi nt - \phi)$, where n is the frequency.)

49. Given a square $ABCD$ of tin with each side equal to $6\sqrt{2}$ ft. A pyramid with square base is to be made as follows: Mark out a square $EFGH$ concentric with the given square, but with EF parallel to the diagonal DB. Then cut out triangles AFB, BGC, CHD, and DEA, and form a pyramid by turning up the 4 equal triangles which are left attached to the square $EFGH$. What is the side of the base for maximum volume of the pyramid?

50. A square sheet of cardboard of side 8 in. is to be used in making a tray, with bottom 4 in. square and uniformly sloping sides, by cutting wedges from the corners of the cardboard and turning up the trapezoidal pieces attached to the 4-inch square center. Determine the side of the top of the tray if its volume is a maximum.

51. Given an isosceles triangle. Inscribe in it the segment of a parabola with its axis bisecting the vertex angle of the triangle such that the area cut from the parabola by the base of the triangle may be a maximum. (*Hint.* Take triangle $(0, 0)$, (a, b), $(a, -b)$.)

52. A rope with a ring in one end is looped over a beam and pulled taut by a heavy weight fastened to the other end and hanging vertically. If the rope slips freely everywhere, and the lower side of the beam is flat, what angle will the rope make with the beam when stretched?

53. A quadrilateral has sides of given length. Prove that for maximum area the quadrilateral can be inscribed in a circle. (*Hint.* Use two opposite angles as parameters; (1) express the area in terms of them, and (2) get the added relation connecting the parameters from the length of the opposite diagonal.)

54. Given a fixed angle BCA. On CA two points D and E are taken, $CD = a$, and $CE = b$. A point F is taken on CB such that the angle EFD is a maximum. How long is CF? (*Hint.* Call $EFD = \theta$. Then $\theta = CFE - CFD$, and $\tan CFD = a\sin C/(x - a\cos C)$, $\tan CFE = b\sin C/(x - b\cos C)$.)

55. Given a point A at a horizontal distance l from another point B. A is to be illuminated by a light L of intensity I, which can move in a groove BC. The angle ABC, or α, is obtuse and lies in a vertical plane. How far up from B must L be located for maximum illumination? Assume the laws that the illumination at a point is directly proportional to the intensity of the source, and to the sine of the angle of incidence, and inversely proportional to the square of the distance from the source. (*Hint.* Call $AL = r$, and angle $LAB = \theta$, and use them as parameters, from which the relation $r = \frac{3}{2} l \cos\theta$ follows for maximum.)

(a) Evaluate the result for $l = 80$ ft., $\alpha = 135°$.

56. The cost of running a boat so as to drive it x miles per hour relative to the water is given by the formula $C = a + bx^3$. It is observed that C is \$44 when $x = 4$ mi. per hour, and \$53½ when $x = 6$ mi. per hour. Calculate the proper speed for most economical operation against a 2-mile current. What is the proper speed when moving with the same current?

57. A battery is to be composed of n cells arranged x in series and y in parallel. The E.M.F. of each cell is e and its resistance r, while the resistance of the external circuit is R. Assuming Ohm's law and the law of divided circuits, show that the current is a maximum when the battery can be so arranged that its internal resistance equals R. (Ohm's law is; current = electromotive force ÷ resistance. For a divided circuit

$$1/r = 1/r_1 + 1/r_2 + \cdots 1/r_y,$$

where r_i is the resistance in the separate branches.)

58. Find the path of a ray of light from a point A in one medium to a point B in another medium, such that a minimum time will be required for light to pass from A to B, the velocity of light in the two media being V_1 and V_2. It is assumed that the required path is in a plane through A and B perpendicular to the plane separating the media.

59. Through a point on the axis of a parabola a chord is drawn which has minimum length. Show that it is perpendicular to the axis.

60. Two sources of heat S_1 and S_2 are at a distance a from each other. The intensities are I_1 and I_2. If the heat at a point is directly proportional to the intensity of the source and inversely proportional to the square of the distance from the source, locate the point O on the line S_1S_2 at which the heat is a minimum.

CHAPTER V

DIFFERENTIALS AND RATES

EXERCISE XXXVI

Differential of Arc and Approximate Length of Arc. Rectangular Coördinates. The differential of arc ds is given by either

$$(1)\ ds = \sqrt{1 + (dy/dx)^2} \cdot dx \quad \text{or} \quad (2)\ ds = \sqrt{1 + (dx/dy)^2} \cdot dy.$$

This fundamental relation between ds, dx, and dy is best remembered by means of a right triangle having dx and dy as its legs and ds as its hypotenuse. The use of (1) or (2), according to convenience, enables us to calculate the approximate length of arc of a curve as follows. Suppose we are considering a curve from $x = a$ to $x = b$; divide the interval from a to b into any number k of equal subdivisions, each of which we call dx. Then knowing also the coördinates of each point of division we can calculate the values of ds at each point. The sum of these k values of ds approximates the length of arc.

Derive the expression for ds in each of the following examples, x being regarded as the independent variable unless otherwise stated. Calculate also approximate length of arc from $x = 0$ to $x = 1$, using intervals of $\frac{1}{3}$ unit, and taking $a = b = 1$ and $c = 0$, wherever they enter. $r = 4$ in example 1.

1. $x^2 + y^2 = r^2$, in terms of dx and in terms of dy.

2. $y^2 = 4\,ax$, in terms of dx and in terms of dy.

3. $y = ax^2 + bx + c$. **4.** $6\,xy = x^4 + 3$. **5.** $y = x^3 - 8\,x$.

6. $y^2 - 4\,x - 4\,y = 0$, regarding y as independent.

7. $x^2/a^2 - y^2/b^2 = 1$. **9.** $y + \log \cos x = 0$.

8. $2\,y + e^x + e^{-x} = 0$. **10.** $y^2 = 2\,x^3 - 6$.

11. $2x^2y - x^2 - 3 = 0$.

12. $y^2 + 3xy = 8$, in terms of dy.

13. $x^2 + 2xy - 4y = 0$.

14. $y^2 = x^3 - 8x$.

15. $9y^2 = (x-2)(x-5)^2$.

16. $y = \sin x$.

The remaining curves are given parametrically and the differential of arc is to be expressed in terms of that parameter. Do not confuse t and τ.

17. $x = t + 1$, $y = t^2$.

18. $x = t^2$, $y = t^3$.

19. $x = 4\sin t$, $y = 4\cos t$.

20. $x = 2\sin t$, $y = 4\cos t$.

21. $x = 2\sin t$, $y = \cos 2t$.

22. $x = 2\sin t$, $y = \sin 2t$.

23. $x = 2\cos t$, $y = \cos 2t$.

24. $x = 3\sin t$, $y = \sin 3t$.

As is obvious from the relation (shown very nicely by a triangle) connecting ds, dx, dy, the sine and the cosine of τ, the inclination of the tangent line to the curve, are given by

$$\sin\tau = dy/ds, \qquad \cos\tau = dx/ds,$$

which gives us an easy method of calculating these functions, particularly for the case of parametric equations. For instance, in Example 17, $ds = \sqrt{1+4t^2}dt$, and $dx = dt$; therefore $\cos\tau = 1/\sqrt{1+4t^2}$.

25. Calculate $\sin\tau$ and $\cos\tau$ for examples 17–24.

These values of $\sin\tau$ and $\cos\tau$ are needed in many problems. For example, if the speed along a path at a given time is v, the components along the X and Y axes respectively are $v_x = v\cos\tau$ and $v_y = v\sin\tau$. This is true whether v is constant or variable. In the next 6 examples calculate v_x and v_y at the points indicated, assuming $v = 4$ units per second. (Substitute numerical values at the earliest possible stage, *always*.)

26. $y = x^2 - 4x$, $x = 0, 1, 2$.

27. $y = x^3 + 1$, $x = 0, 1$.

28. $y^2 = x^3 - 4x$, $x = 1, 2$.

29. $y = 2x - x^2$, $x = 1, 2$.

30. $x^2 + y^2 = 6x$, $x = 0, 2, 3$.

31. $xy = x^2 + 4$, $x = 1$.

32. The equation of the path of a projectile fired at an angle of elevation a, with an initial velocity v_0, is $y = x\tan a - (gx^2)/(2v_0^2\cos^2 a)$, where g is a constant and the starting point is the origin. Assuming

$a = 45°$, $v_0 = 64$, and $g = 32$, find the vertical and horizontal components of the velocity at the positions indicated:

(a) When it has gone 32 ft. horizontally. (Its velocity then is $16\sqrt{10}$.)
(b) When it is at a vertical height of 32 ft. (Velocity then is $32\sqrt{2}$.)
(c) When it is 24 ft. vertically above the ground.

EXERCISE XXXVII

Differential of Arc. Polar Coördinates. Here again we have a choice of two formulas, according as θ or ρ is taken as the independent variable.

$$(1)\ ds = \sqrt{\rho^2 + (d\rho/d\theta)^2} \cdot d\theta \quad \text{or} \quad (2)\ ds = \sqrt{\rho^2(d\theta/d\rho)^2 + 1} \cdot d\rho.$$

We rarely use (2) in applications. The relation connecting $d\rho$, $d\theta$, and ds is easily remembered by means of a right triangle with $d\rho$ and $\rho d\theta$ as the two legs and ds as the hypotenuse. Find ds for the following in form (1):

1. $\rho = a \cos\theta$	7. $\rho = 4\sin^3\theta/3$.	13. $\rho = 3/(2 - \cos\theta)$.
2. $\rho = b \sin\theta$.	8. $\rho = \log\sin 3\theta$.	14. $\rho = 3/(1 - 2\cos\theta)$.
3. $\rho = 4\sin\theta + 3\cos\theta$.	9. $\rho = \sin^2\theta$.	15. $\rho = \log\tan\theta$.
4. $\rho = 5\cos\theta - 12\sin\theta$.	10. $\rho = 4\sin 2\theta$.	16. $\rho = \sin\theta\tan\theta$.
5. $\rho = 1 + \cos\theta$.	11. $\rho = 4\tan\theta/4$.	17. $\rho = \sec^2\theta/2$.
6. $\rho = \sin 2\theta + \cos 2\theta$.	12. $\rho = 3/(1 + \cos\theta)$.	18. $\rho = 2 - 3\sin\theta$.

It is sometimes desirable to know $\sin\psi$ and $\cos\psi$. Here again from the triangle representation we see that $\sin\psi = \rho d\theta/ds$ and $\cos\psi = d\rho/ds$.

19. Calculate $\sin\psi$ and $\cos\psi$ for each of the first 12 examples in the preceding list.

EXERCISE XXXVIII

Differentials. While no new power in the process of differentiating is gained by the use of differentials, it is well for the student to become familiar with this notation. In the following examples write down dy. In some cases it would be well to write down the second or higher differential.

1. $y = \frac{1}{2}\sin x + \sin\frac{1}{2}x$.	3. $y = x^2\sqrt{2 - x}$.	5. $y = \sqrt{4x + x^2}$.
2. $y = 2/\sqrt[3]{x} - 3x^{\frac{2}{3}} - 5/x$.	4. $y = (x^2 - \sqrt{x})^5$.	6. $y = \sqrt{1 - x^3}$.

7. $y = 1/(2 - 3x^2)^3$.

8. $y = x/\sqrt{1 + x}$.

9. $y = (x + 1/x)(3\sqrt{x} - 5x^3)$.

10. $y = (x + \sqrt{x})/(x - \sqrt{x})$.

11. $y = x/\sqrt{x^2 - 5}$.

12. $y = \cos^2 a/x$.

13. $y = \text{arc sin}\sqrt{1 - x^2}$

14. $y = \log(x^2 + \sqrt{1 + x^2})$.

15. $y = \sqrt{2x - x^2} - 2\tan^{-1}\sqrt{(2 - x)/x}$.

16. $y = \log_5\sqrt{1 - 3x^2}$.

17. $y = \log_4^3[(1 - x)/(1 + x)]$.

18. $y = (x^2 - 2x - 2)e^x$.

19. $y = e^x/x^3$.

20. $y = e^{\sqrt[3]{x}}$.

21. $y = e^x(x - 2)/x^2$.

22. $y = 5^{x^2}$.

23. $y = \log(2 - \sqrt{x})$.

24. $y = \sin\sqrt{1 - x}$.

25. $y = \sec a^x$.

26. $y = \cos^3 x^2$.

27. $y = x^{\sqrt{1 - x^2}}$.

28. $x^2 + 9y^2 = 9$.

29. $2x^2 + y^2 - 2y + x - 2 = 0$.

30. $x^3 - x^2y + y^3 = 0$.

EXERCISE XXXIX

Change of Variable. (A) ***Change of Independent Variable from x to y.*** This depends on the simple relation

$$dy/dx = 1 \div dx/dy,$$

from which the equalities given below are derived by simple differentiation:

$$\frac{d^2y}{dx^2} = -\frac{d^2x}{dy^2}\cdot\left(\frac{dx}{dy}\right)^{-3}, \quad \frac{d^3y}{dx^3} = -\left[\frac{d^3x}{dy^3}\cdot\frac{dx}{dy} - 3\left(\frac{d^2x}{dy^2}\right)^2\right]\left(\frac{dx}{dy}\right)^{-5}.$$

Transform the following:

1. $x\dfrac{d^2y}{dx^2} + y\dfrac{dy}{dx} = 0.$

2. $\dfrac{dy}{dx} + \left(\dfrac{dy}{dx}\right)^3 + (y - 2)\dfrac{d^2y}{dx^2} = 0.$

3. $2\left(\dfrac{d^2y}{dx^2}\right)^2 - 5\dfrac{dy}{dx}\cdot\dfrac{d^3y}{dx^3} + 3\dfrac{d^2y}{dx^2}\cdot\left(\dfrac{dy}{dx}\right)^2 = 0.$

4. $\left(3\dfrac{dy}{dx} + 2\right)\left(\dfrac{d^2y}{dx^2}\right)^2 - \left(2\dfrac{dy}{dx} + 1\right)\dfrac{dy}{dx}\cdot\dfrac{d^3y}{dx^3} = 0.$

5. $(y - 4)\left(\dfrac{dy}{dx}\right)^3 + \dfrac{dy}{dx} - \dfrac{d^2y}{dx^2} = 0.$

6. $\left(2\dfrac{dy}{dx} + 3\right)\left(\dfrac{d^2y}{dx^2}\right)^2 + \left(\dfrac{dy}{dx} + 4\right)\dfrac{dy}{dx}\cdot\dfrac{d^3y}{dx^3} = 0.$

7. $xy\dfrac{d^3y}{dx^3} + y^2\dfrac{dy}{dx}\cdot\dfrac{d^2y}{dx^2} = \left(\dfrac{dy}{dx}\right)^4.$

8. $\dfrac{d^2y}{dx^2}\cdot\dfrac{d^3y}{dx^3} = y\left(\dfrac{dy}{dx}\right)^4.$

(B) ***Change of Dependent Variable.*** Here the relation between the old dependent variable, y, and the new one, z, is given. Call it $y = \phi(z)$. The formulas of transformation are then

$$dy/dx = \phi'(z)\,dz/dx, \qquad d^2y/dx^2 = \phi''(z)(dz/dx)^2 + \phi'(z)\,d^2z/dx^2.$$

Formulas for higher derivatives are easily calculated. In the following examples change the dependent variable:

9. $\dfrac{x}{2}\dfrac{d^2y}{dx^2} - \dfrac{x}{4y^2}\left(\dfrac{dy}{dx}\right)^2 + \dfrac{1}{2}\dfrac{dy}{dx} - y = 0,\ y = z^2.$

10. $\dfrac{d^2y}{dx^2} + \dfrac{y}{1-y^2}\left(\dfrac{dy}{dx}\right)^2 + 2\dfrac{dy}{dx} + 2y^2 - 2 = 0,\ y = \sin z.$

11. $(1+y)^2\dfrac{d^2y}{dx^2} - 2y\left(\dfrac{dy}{dx}\right)^2 - 2(1+y^2)\dfrac{dy}{dx} = 0,\ y = \tan z.$

12. $\dfrac{y}{1-y^2}\cdot\dfrac{d^2y}{dx^2} + \dfrac{1}{y^2}\left(\dfrac{dy}{dx}\right)^2 = 0,\ y = \sin z.$

13. $y^3\dfrac{d^3y}{dx^3} - 5y^2\dfrac{d^2y}{dx^2} + 11y\dfrac{dy}{dx} - 5 = 0,\ y = e^z.$

14. $\dfrac{d^3y}{dx^3} + y\dfrac{d^2y}{dx^2} - y^2\dfrac{dy}{dx} - 10y^3 = 0,\ y = 1/z.$

(C) ***Change of the Independent Variable.*** Here we introduce a new independent variable, the defining relation being given. The procedure here is obvious, as $dy/dx = dy/dt \cdot dt/dx$.

15. $x^2\dfrac{d^2u}{dx^2} + x\dfrac{du}{dx} + u = 0,\ y = \log x.$

16. $y + \dfrac{1}{x}\dfrac{dy}{dx} + \dfrac{d^2y}{dx^2} = 0,\ x^2 = 4t.$

17. $(1-y^2)\dfrac{d^2u}{dy^2} - y\dfrac{du}{dy} + 4u = 0,\ x = \operatorname{arc}\cos y.$

18. $(x - x^3)\dfrac{d^2y}{dx^2} + (1 - 3x^2)\dfrac{dy}{dx} - xy = 0,\ x = \sqrt{1-t^2}.$

19. $\dfrac{d^2y}{dx^2} + \dfrac{2x}{1+x^2}\dfrac{dy}{dx} + \dfrac{y}{(1+x^2)^2} = 0,\ x = \tan z.$

20. $\dfrac{d^2y}{dx^2} + \dfrac{1}{x}\dfrac{dy}{dx} + y = 0,\ x^2 = 4t.$

21. $(1+x^2)\dfrac{d^2y}{dx^2} - x\dfrac{dy}{dx} - y = 0,\ x = \sin z.$

22. $(1-x^2)\frac{d^2y}{dx^2}-x\frac{dy}{dx}-2x=0,\ x=\sin t.$

23. $(1+x^2)\frac{d^2y}{dx^2}+2x\frac{dy}{dx}+4=0,\ x=\tan t.$

24. $x^4\frac{d^2y}{dx^2}+x^3\frac{dy}{dx}+3y=0,\ x=\frac{1}{t}.$

25. $\frac{1-x^2}{x^2}\frac{d^2y}{dx^2}-\frac{1}{x^3}\frac{dy}{dx}+y=0,\ x=\sqrt{1-t^2}.$

26. $(1-x^2)^2\left(\frac{d^2y}{dx^2}\right)^2-2x(1-x^2)\frac{d^2y}{dx^2}\cdot\frac{dy}{dx}+\left(\frac{dy}{dx}\right)+1=0,$ $x=\sin t.$

27. $\frac{d^2y}{dx^2}+\frac{2x}{1+x^2}\frac{dy}{dx}+\frac{y}{1+x^2}=0,\ x=\tan t.$

28. $\sqrt{1-x^2}\frac{d^3y}{dx^3}-y\frac{dy}{dx}=0,\ x=\sin z.$

EXERCISE XL

Rate Problems. In most of these problems time, denoted by t, is the independent variable. Hence if x is any quantity which changes with the time, dx/dt is the time rate of change of x. In general, dy/dx may be defined as a rate — the relative rate of change of y with respect to x. If several dependent quantities are changing simultaneously, it is necessary first to get an equation (or equations) connecting them. Then differentiating with respect to t gives a relation (or relations) connecting the several time rates. Substitute in this (or these) the particular values known at the time under consideration. Solve the resulting equation (or equations) for the unknown rate or rates. *Do not substitute a constant value* for any quantity which varies with the time *until after differentiation.*

1. A stone is dropped into a calm lake, sending out a series of concentric ripples. If the radius of the outer ripple increases uniformly 6 ft./sec., how rapidly is the disturbed area increasing at the end of 2 sec.?

2. A rope ABC, 28 ft. long, runs over a pulley B, which is 12 ft. above the level track on which the ends A and C are moving. If the rope is taut, and the end C moves to the right 13 ft./sec., how fast will A move when C is 5 ft. from E, a point on AC directly below B?

3. A right circular cone keeps its volume constant. If the radius of the base is increasing at the rate of 2 in./sec., how fast is the altitude changing when the altitude is 6 in. and the radius 4 in.?

4. A right prism has constant altitude, 4 ft. If its base is an equilateral triangle of side 2 ft., and the sides are increasing 1 ft. per hour, how fast is the volume increasing? How fast the surface?

5. A boat is fastened to a rope which is wound about a windlass 24 ft. above the level at which the rope is attached to the boat. The boat is drifting away at the rate of 10 ft./sec. How fast is it unwinding the rope when 32 ft. from the point directly under the windlass? What is the acceleration of a point on the rope?

6. A bundle is being elevated by a rope passing over a pulley 25 ft. above the ground. A man takes hold of a ring in the extreme end of the rope, which is 50 ft. long, and walks away 10 ft./sec. How rapidly will the bundle ascend at the start? The man's hands are assumed to be 5 ft. above the ground.

7. A cistern is in the shape of an inverted circular cone, with its diameter equal to its height, each being 10 ft. How fast is the water pouring in when it is 5 ft. deep and rising 4 in./min.?

8. There is a cistern in the shape of a cone 12 ft. deep and 6 ft. in diameter at the top, into which 10 cu. ft. of water are pouring each minute. If at the time the water is 8 ft. deep it is observed to be rising only 6 in./min., how much water is leaking away?

9. Sand is being poured on the ground, forming a conical pile with its altitude equal to $\frac{2}{3}$ of the radius of its base. If the sand is falling at the rate of 12 cu. ft./sec., how fast is the altitude increasing at the time it is 5 ft.?

10. A balloon is in the shape of an ellipsoid of revolution, its longer axis being twice the two shorter ones. If gas is being poured in at the rate of 100 cu. ft./min., when the balloon is 10 ft. long and expanding uniformly how fast is the length increasing?

11. A right prism has an equilateral triangle for a base. How rapidly are its volume and surface changing when the side of the base is 3 in. and increasing 2 in./min., while the altitude is 5 in. and decreasing 3 in./min.?

12. A right circular cylinder is capped by a hemispherical top. If the radius of the common base increases at the rate of 2 in./sec. and the altitude of the cylinder 3 in./sec., at what rate is the volume increasing when the radius is 5 and the altitude 10?

13. A right prism has square bases. If the sides of the bases are 2 in. and increasing 1 in./sec., while the altitude is 4 in. and decreasing 2 in./sec., (a) how fast is the volume increasing? (b) When will the volume cease increasing?

14. Two ships are at the same point. The first leaves at 10 A.M., sailing east at the rate of 9 mi./hr. The second starts at 11 A.M. and sails south 12 mi./hr. How fast are they separating at noon?

15. A man is walking 6 ft./sec. across a bridge 60 ft. above the water level below. A man in a rowboat is rowing at right angles to the bridge 4 ft./sec. How rapidly are the men separating at the end of 5 sec.?

16. A rod 10 ft. long moves so that its ends A and B remain constantly in the x- and the y- axis respectively. If A is 8 ft. from the origin and is moving away at the rate of 2 ft./sec.,

(a) at what rate is the end B coming down?
(b) at what rate is the area formed by AB and the axes changing?
(c) if P is the middle point of AB, at what rate is OP changing?

17. The side in each of the following regular polygons is 12 in. and expanding 4 in. per hour. How fast is the area increasing for (a) a square? (b) a triangle? (c) a hexagon? (d) an n-sided polygon?

18. The hour hand of a clock is 3 ft. long and the minute hand is 4 ft. long. At what rate are the ends of the hands approaching each other at (a) 3 o'clock? (b) 4 o'clock?

19. An angle θ is increasing uniformly. Show that (a) when $\theta = 0$ the sine and tangent are increasing at the same rate; (b) when $\theta = \pi/6$ the cosine is decreasing $\frac{3}{8}$ as fast as the tangent is increasing; (c) when $\theta = \pi/4$ the tangent is increasing $\sqrt{2}$ times as fast as the secant, while the sine is increasing at the same rate the cosine is decreasing; (d) when $\theta = \pi/3$ the tangent is increasing 8 times as fast as the sine, while the secant is increasing 4 times as fast as the cosine is decreasing.

20. A man is crossing at right angles a street 60 ft. wide. There is a lamp on the opposite side, which is 10 ft. upstreet from the point which he is approaching. The lamp is on a level with the man's head, and casts his shadow against the buildings on the side he is leaving. If the man walks 10 ft./sec., what is the rate of his shadow on the buildings when he is $\frac{3}{4}$ the way across?

21. The adiabatic law for the expansion of air is $PV^{1.4} = C$. If at a given time the volume V is observed to be 10 cu. ft. and the pressure is 50 lb. per square inch, at what rate is the pressure changing if the volume is decreasing 1 cu. ft. per second?

22. A kite is at an altitude of 160 ft. and there are 280 ft. of cord out. If the kite is moving horizontally 4 mi./hr. directly away from the person flying it, how fast is the cord being paid out?

23. Two sides of a triangle increase at the rate of 2 in. per second and the included angle at the rate of $\frac{1}{5}$ radian per second. At what rate is the area changing at the instant when each of the three sides is 10 in. long?

24. A revolving light is $\frac{1}{2}$ mi. from a straight beach. A ray from the light illumines a spot on the beach. If the light makes a complete revolution per minute, how fast is the spot moving along the beach when the ray makes a 45° angle with the shore line? How fast at the end of 10 sec. from the time it is directed toward the nearest point on the shore?

25. I have a variable right triangle ABC, with B, the vertex of the right angle, moving along the axis of the parabola $y^2 = 4x - 4$, C being fixed at the vertex of the parabola, and A tracing out the upper half of the parabola. If B starts at the vertex and recedes uniformly 2 in./sec., how fast is A receding from the origin of coördinates at the end of 2 sec.? How fast is the area of the triangle changing?

26. The strength of a beam is given by the formula $S = kx^3y$, where k is a constant of the material, y is the width and x the thickness of the beam. If at a given time $x = 12$ in. and decreases 1 in./min., (a) how fast should y, which is 4 in., be increasing if the strength is to remain constant? (b) How fast is the area of the cross section changing?

27. A wheel of radius 5 ft. is turned by a connecting shaft 13 ft. long, fastened to a piston. If the wheel turns uniformly at the rate of 100 rev./min., (a) at what rate is the piston moving when the connection on the rim of the wheel is at its highest and lowest points? (b) At what position has the piston maximum velocity? (c) What is the angular velocity of the wheel at each of these times?

28. If in Ex. 27 the radius of the wheel is 4 ft. and the connecting rod is 6 ft., what is the rate of the piston when the line joining the connection to the center O of the wheel makes a 60° angle with OP, P being the center of the piston head? At what points does the velocity of the piston become zero?

29. A grade crossing has a double gate, the two arms of which rotate upward about the same axis. The arm over the driveway is 12 ft. while that over the sidewalk is but 5 ft. Each arm rotates at the rate of 4 radians per minute. At what rate is the distance between their extremities changing (a) when each makes a 45° angle with the horizontal? (b) when each makes a 60° angle?

30. A man is walking over a bridge at the rate of 3 mi. per hour, and a boat passes under the bridge at right angles, going 7 mi. per hour. The bridge is 18 ft. above the boat. How fast are the boat and the man separating (a) 1 min. later? (b) 4 min. later?

31. If the x intercept of the tangent to the positive branch of the hyperbola $xy = 4$ is increasing 3 units per second, find the rate and acceleration with which the y intercept is changing at the end of 5 sec., the x intercept starting from the origin.

32. If the x intercept of the tangent to the curve $y = e^{-x}$ is increasing constantly 5 units per second, find the velocity and acceleration of the y intercept at the instant when the x intercept is 12 units.

33. Given an inclined plane BC, with a 30° angle of elevation. At B, the top of the plane, is a perpendicular post BA, 10 ft. long, with a pulley at the top A. A mass M_1, attached to a rope, starts at B and is moved down the plane with a uniform velocity of 10 ft./sec. The rope passes over the pulley A and supports another mass M_2. How fast is M_2 going up at the end of 1 sec.? at the end of 2 sec.?

34. A reservoir is in the shape of a frustum of a right circular cone with lower base of radius 10 ft., upper base 15 ft., and depth 10 ft. How fast are the volume of water and the area of the surface of the water increasing when the water is 5 ft. deep and rising 1 ft. per hour?

35. A right triangular lot ABC has $AB = 5$ rd. and $BC = 12$ rd. There is a street lamp at A. A man walks at the uniform rate of 13 ft. per second from B to E, the mid-point of the hypotenuse AC. (a) How fast is his shadow moving when it is halfway from B to C? (b) How fast when the man is halfway from B to E? (c) when he reaches E? (d) What is the acceleration of the shadow when it is halfway from B to C?

36. A boy is whirling a stone on the end of a string with an angular velocity of 4 radians per second and is allowing the string to slip through his hand at the rate of 5 ft./sec. At the time the stone is 3 ft. from his hand, at what rate is it moving? What is the component of its velocity at right angles to the string?

37. A moving point traces out each of the following curves. Find the rate at which the arc is increasing in each case under the given conditions:

(a) $y^2 = 2x + 5$, $x = 2$, $dx/dt = 3$.
(b) $3x^2 + 4y^2 - 12 = 0$, $y = \frac{3}{2}$, $dy/dt = 1$.
(c) $2x^2 - 3y^2 - 2 = 0$, $y = 4$, $dx/dt = 4$.
(d) $y = x^2 - 2x + 3$, $x = 2$, $dx/dt = 4$.
(e) $y = x^3 - x^2$, $x = 2$, $dy/dt = 3$.

38. A point P moves on the curve $y = 2x^3$ at the uniform rate of 4 ft. per second. At what rate is the inclination of the tangent to the curve at P changing as P passes through the point $(1, 2)$?

39. A military observer in an aëroplane is ascending at the rate of 12 mi. per hour. How fast is the visible area of the earth's surface increasing in square miles per minute at the end of 5 min.? at the end of 10 min.?

40. The ends of a taut cord are A and B. The extremity A moves along the line $y = 3$, with the velocity of 2 units per second. The extremity B is attached to a bead which is free to move on the negative x-axis. A portion of the cord between A and B passes over the circle $x^2 + y^2 = 9$. Find the velocity of B when the bead passes through $(-5, 0)$.

41. A candle is moving directly away from the center of a sphere of radius 5 ft. at the rate of 2 ft. per second. At what rate is the surface illuminated by the candle increasing when the candle is 10 ft. from the center of the sphere?

42. A cylinder with an open top, a height of 4 ft., and a base of radius 1 ft. is filled with water. It is then tipped around a tangent to the base at a uniform rate of 10° per second. At what rate is the water overflowing at the end of 3 sec., assuming that the surface of the water is always horizontal?

43. One end A of a rod AB slides on the x-axis while the rod passes over and rests on the rim of a wheel of radius 5 in., with center at the origin O. If AB is 26 in. long and A moves 6 in. per second, at what rate is B approaching O when OA is 13 in.? What is the velocity of B at this time?

44. If the radius of a sphere is 12 in. and is increasing at the rate of $\frac{1}{12}$ in. per second, at what rate is the volume of the maximum inscribed right cylinder increasing?

45. A horizontal cylindrical tank turns on its axis and empties through a slit cut along an element. The level of the fluid in the cylinder remains the same as the level of the slit. If the slit revolves uniformly 10° per second, at what rate is the tank emptying when the slit has turned through 30° from its extreme vertical position?

In many problems the relative rate of increase or the percentage rate of increase is the important idea. The relative and percentage rates should be calculated for a number of the preceding examples. If dy/dt is the time rate of increase with respect to t, $dy/dt \div y$ is the *relative rate of increase* of y, and

100 times this is the *percentage rate of increase.* Note that the relative rate is therefore simply the logarithmic derivative of y with respect to t. Note also generally that dy/dx is an absolute rate of change in y with respect to x, while $dy/dx \div y$ is the relative rate.

46. If the length and breadth of a rectangle each increase $\frac{1}{10}\%$ per degree as a metal plate is heated, show that the area increases $\frac{1}{5}\%$ per degree.

47. A circular hill has a central vertical section of the form $x^2 + 160y = 1600$ where the unit is 1 yd. If the top is being cut down in horizontal layers, at what relative rate per vertical yard is the area of the section increasing after 4 yd. have been removed?

48. Assume the law $pv^{1.4} = k$ for air under pressure. If the volume is changing, what is the relative change in pressure per unit increase in volume when the volume is 28 cubic units? What is the percentage change?

49. The deflection of a uniformly loaded beam supported at one end is given by $y = k(\frac{1}{2}l^2x^2 - \frac{1}{3}lx^3 + \frac{1}{12}x^4)$, where l is the total length of beam and x the distance from the point of support. What is the relative change of the deflection at the middle of the beam per unit increase in x? at the free end?

50. Van der Waals' equation for a gas is $(p + a/v^2)(v - b) = k$, where p is the pressure, v is the volume, and a, b, and k are constants. What is the change in volume per unit increase in pressure? What is the relative change?

CHAPTER VI

INDETERMINATE FORMS

EXERCISE XLI

Indeterminate Forms. $0 \div 0$. If a fraction $f(x)/F(x)$, where numerator and denominator are both functions of x, assumes the indeterminate form $0/0$ for a particular value of x, its limiting value may be found by differentiating $f(x)$ for a new numerator, and $F(x)$ for a new denominator, giving $f'(x)/F'(x)$. If this form is still indeterminate, repeat the process. In the case of trigonometric functions it is frequently desirable or necessary to change the form of the functions before differentiating. Note that you do *not* differentiate the fraction. Evaluate each of the following for the value of x given. (Assure yourself first that the forms are indeterminate.)

1. $\dfrac{xe^{3x} - x}{1 - \cos 2x}$, $x = 0$. **2.** $\dfrac{e^x - 1 - x}{\cos 2x - 1}$, $x = 0$. **3.** $\dfrac{6x^2 - 5x + 1}{8x^2 - 2x - 1}$, $x = \dfrac{1}{2}$.

4. $\dfrac{x^3 - 2x^2 - x + 2}{x^3 - 7x + 6}$, $x = 1, 2$.

5. $\dfrac{x^3 + 8}{x^5 + 32}$, $x = -2$.

6. $\dfrac{2 - \sqrt{x - 3}}{x^2 - 49}$, $x = 7$.

7. $\dfrac{e^x - \sin 2x - e^{-x}}{x - \sin x}$, $x = 0$.

8. $\dfrac{e^x - e^{-x} - 2\sin x}{4x^3}$, $x = 0$.

9. $\dfrac{1 - \cos x - x \sin x}{2 - 2\cos x - \sin^2 x}$, $x = 0$.

10. $\dfrac{x^2 - x\sin x}{e^x + e^{-x} - x^2 - 2}$, $x = 0$.

11. $\dfrac{\log x - \sin(x - 1)}{(x - 1)^2}$, $x = 1$.

12. $\dfrac{\sin 4x + e^{-2x} - 2x^2 - 2x - 1}{x^4 + 12x^3}$, $x = 0$.

13. $\dfrac{(2 - x^2)\sin x - 2x\cos x}{x^3}$, $x = 0$.

14. $\dfrac{\sqrt{3x} - \sqrt{12 - x}}{2x - 3\sqrt{19 - 5x}}$, $x = 3$.

15. $\dfrac{\sqrt{16x - x^4} - 2\sqrt[3]{4x}}{2 - \sqrt[4]{2x^3}}$, $x = 2$.

Find the slopes of the following curves at the points indicated:

16. $a^2y^2 - a^2x^2 - x^4 = 0$, at (0, 0).

17. $y^3 = ax^2 - x^3$, at (0, 0).

18. $x^3 - 3\,axy + y^2 = 0$, at (0, 0).

19. $x^4 - a^2xy + b^2y^2 = 0$, at (0, 0).

20. $(y^2 + x^2)^2 - 6\,axy^2 - 2\,ax^3 + a^2x^2 = 0$, at (0, 0) and $(a, 0)$.

21. Given a circle with center O and horizontal radius OA. AP is the tangent at A. A variable point M starts at A and moves about on the circle; AP is laid off on the tangent $= AM$, the arc; and PM, a straight line, is drawn cutting AO (extended) at B. If we call the variable angle $AOM = \theta$, what is the limiting value of OB as θ approaches zero?

EXERCISE XLII

Indeterminate Forms. $\infty \div \infty$. In case a certain value of x makes both the numerator and the denominator infinite, the value of the fraction becomes indeterminate also. The limiting value is obtained in precisely the same manner as in the preceding exercise. The student should recall that it is sometimes necessary to change the form of the function before evaluating, as otherwise the value continues indeterminate indefinitely.

1. $\dfrac{\log \sin 2\,x}{\log \sin x}$, $x = 0$.

2. $\dfrac{\cot x}{\cot 3\,x}$, $x = 0$.

3. $\dfrac{x^5}{e^x}$, $x = \infty$.

4. $\dfrac{\tan x}{\tan 3\,x}$, $x = \pi/2$.

5. $\dfrac{\cot 2\,x}{\cot x}$, $x = 0$.

6. $\dfrac{e^x}{x^3}$, $x = \infty$.

7. $\dfrac{\log x}{a^x}$, $x = \infty$.

8. $\dfrac{\log \tan x}{\log \tan 2\,x}$, $x = \pi/2$.

9. $\dfrac{\pi/x}{\cot \pi x/2}$, $x = 0$.

10. $\dfrac{x^{-1}}{\cot x}$, $x = 0$.

11. $\dfrac{\log(1 - x)}{\cot \pi x}$, $x = 1$.

12. $\dfrac{\sec \pi x}{\tan 3\,\pi x}$, $x = \dfrac{1}{2}$.

13. $\dfrac{x^4 + x^3}{e^x + 1}$, $x = \infty$.

14. $\dfrac{\log(a + x)}{x}$, $x = \infty$.

15. $\dfrac{\sec \pi x/2}{\log \sec \pi x/2}$, $x = 1$.

16. $\dfrac{\log 1/x}{\cot x}$, $x = 0$.

17. $\dfrac{x + \log x}{x \log x}$, $x = \infty$.

18. $\dfrac{\log(x - 1) + \tan \pi x/2}{\cot \pi x}$, $x = 1$.

EXERCISE XLIII

Indeterminate Forms. $0 \cdot \infty$, $\infty - \infty$. If a product contains one factor equal to 0 and another equal to infinity for a certain value of the independent variable, the value is again indeterminate. But in this case the form can be rewritten so as to assume the form $0 \div 0$ or $\infty \div \infty$, since $0 \cdot \infty = 0 \div 1/\infty = 0 \div 0$. Similarly, an indeterminate form of the type $\infty - \infty$ can be changed into the form $0 \div 0$ or $\infty \div \infty$ and then evaluated. Evaluate:

1. $x \log \sin x$, $x = 0$.
2. $\sin x (\log x)$, $x = 0$.
3. $\csc x \sin (\tan x)$, $x = 0$.
4. $(2 - x) \tan \pi x/4$, $x = 2$.
5. $(1 - x) \tan \pi x/2$, $x = 1$.
6. $(1 - \sin x) \tan x$, $x = \pi/2$.
7. $2/(x^2 - 1) - 1/(x - 1)$, $x = 1$.
8. $x/(x - 1) - 1/\log x$, $x = 1$.
9. $1/x^2 - 1/(x \tan x)$, $x = 0$.
10. $x \tan x - \frac{1}{2} \pi \sec x$, $x = \pi/2$.
11. $\dfrac{1}{\sin^2 x} - \dfrac{1}{x^2}$, $x = 0$.
12. $\dfrac{x^2 - 4}{x^2} \tan \dfrac{\pi x}{4}$, $x = 2$.
13. $\dfrac{x - 1}{2x^2} - \dfrac{1}{x(e^{2x} - 1)}$, $x = 0$.
14. $\dfrac{1}{\log (1 + x)} - \dfrac{1}{x}$, $x = 0$.
15. $\dfrac{1}{\sin^3 x} - \dfrac{1}{x^3}$, $x = 0$.
16. $\dfrac{1}{1 - x} - \dfrac{14}{5 - x^2 - 4x^3}$, $x = 1$.
17. $2^x \tan \pi/2^x$, $x = \infty$.
18. $x \log (1 + 1/x)$, $x = \infty$.
19. $\sec x - 1/(1 - \sin x)$, $x = \pi/2$.
20. $\sin x (\log x)^2$, $x = 0$.

EXERCISE XLIV

Other Indeterminate Forms. Certain values of the independent variable may cause a given function to assume one of the following indeterminate forms, 0^0, 1^∞, ∞^0. These are all evaluated by taking the logarithm of both sides and then evaluating the right-hand member, which will be of the form $0 \cdot \infty$ for the value of x which made the original function indeterminate. Evaluating this gives the value of the logarithm of the function. $(1 + 1/x)^x$, as $x \doteq \infty$, is a well-known illustration. Evaluate:

1. $y = (\sin x)^{\sec x}$, $x = \pi/2$.
2. $y = (\sin x)^{\tan x}$, $x = \pi/2$.
3. $f(x) = (1 + 2x)^{1/x}$, $x = 0$.
4. $f(x) = (1/x)^{\sin x}$, $x = 0$.

5. $f(x) = (1 - x)^{\tan \pi x}$, $x = 1$.

6. $f(x) = (2 - x)^{\frac{1}{2} \tan \pi x}$, $x = 1$.

7. $f(x) = [(\tan x)/x]^{1/x}$, $x = 0$.

8. $f(x) = (1/x)^{\tan x}$, $x = 0$.

9. $f(x) = (\cos 2/x)^{x}$, $x = \infty$.

10. $f(x) = (\cos 2/x)^{x^2}$, $x = \infty$.

11. $f(x) = (\cos 2/x)^{x^3}$, $x = \infty$.

12. $f(x) = (x + 1)^{\cot x}$, $x = 0$.

13. $f(x) = (1 + 3x)^{(1-2x)/x}$, $x = 0$.

14. $f(x) = (1 - x/2)^{\frac{1}{2} \tan \pi x}$, $x = 2$.

15. $y = (1 + x)^{\log x}$, $x = 0$.

16. $y = x^{1/(1-x)}$, $x = 1$.

17. $y = (e^{2x} + 2x)^{1/4x}$, $x = 0$.

18. $y = (\log x)^{1/(1 - \log x)}$, $x = e$.

CHAPTER VII

CURVATURE. EVOLUTES

EXERCISE XLV

Curvature and Radius of Curvature. If $R =$ radius of curvature and $K =$ curvature, the formulas to be used are

$$(1)\quad R = [1 + (dy/dx)^2]^{3/2}/(d^2y/dx^2),$$
$$(2)\quad R = [1 + (dx/dy)^2]^{3/2}/(d^2x/dy^2),$$
$$(3)\quad K = 1/R.$$

If the curve is defined parametrically, be careful in finding d^2y/dx^2, as in Exercise XXXI. In the following examples calculate R and find the equation of the normal. Then draw both the curve and the normal, and lay off a length R along the normal on the concave side of the curve. This point is called the center of curvature. The pupil *should note that* (2) *is an alternative form which is very convenient when* $dy/dx = \infty$, *and at other times.*

1. $y = x^2/2$, where $x = 1$.

2. $y = x^3 - 1$, at $x = 1$.

3. $6y = x^3 - 12x - 2$, at $(2, -3)$.

4. $x^2 - 4y^2 = 12$, at $(4, 1)$.

5. $y^2 = 10x - 6$, at $(1, 2)$.

6. $y^2 = 8 - 4x$, at $(1, 2)$.

7. $x^2 + 4y^2 - 10x = 0$, at $(2, 2)$.

8. $x^2 - 4y^2 + 6x = 0$, at $(2, 2)$.

9. $y^2 = x^3 - 4x^2 + 4x$, at $(1, 1)$.

10. $y^2 = x^3 + 8$, at $(-2, 0)$.

11. $y = \cos(x - 1)$, at $(1, 1)$.

12. $x^2 - 2xy + 2y^2 + x = 0$, at $(-1, 0)$.

13. $x = 3t$, $y = 2t^2 - 1$, at $t = 1$.

14. $x = 4t$, $y = 2/t$, at $t = 1$.

15. $x = t^2 - 2$, $y = t - 1$, at $t = 1$.

16. $x = 2t$, $y = 2t^2 - 1$, at $t = 1$.

17. $x = t^2 - 2t - 3$, $y = t^3 - 4t - 4$, at $t = 2$.

18. $x = t^2$, $y = t^3 - 4t$, at $t = -1$.

19. $x = 4\sin t$, $y = 2\cos t$, where $x = 2$.

20. $x = 2\cos t$, $y = \cos 2t$, at $t = \pi/2$.

In the following curves calculate K and R, and draw the figure, with the circle of curvature at the given point:

21. $y = e^x$, where it crosses the y-axis.

22. $y = e^{-x^2}$, where it crosses the y-axis.

23. $y = x^4 - 4x^3 - 18x^2$, at the origin.

24. $x^3 + xy^2 - 6y^2 = 0$, at $(3, 3)$.

25. $x^2 + 2y^2 - 2xy - x = 0$, at $(1, 0)$.

26. $y^2(x - 4m) = x(x - 3m)$, at $x = 0$ and $x = 3m$.

27. $e^x = \sin y$, at $(0, \pi/2)$.

28. $xy = a^2/2$, at the point (x, y).

29. Find the point on $y = \log x$ where K is a maximum.

Find the points of maximum and minimum curvature on the curves of examples 30–35, and calculate K for three points.

30. $y = \sin x$.

31. $b^2x^2 + a^2y^2 = a^2b^2$; $a > b$.

32. $y = x^3 - 6x^2 + 9x$.

33. $y = ax^2 + bx + c$.

34. $x^{1/2} + y^{1/2} = 1$. Also find R at $(1, 0)$.

35. $y = x^2 - 4x + 2$. What is the nature of the point where the curvature is a maximum?

36. Find the locus of points at which the curvature of the family of curves $y = x^n$, $x > 0$, $n > 0$ is a maximum.

EXERCISE XLVI

Center of Curvature and Evolutes (Cartesian). In the preceding exercise we constructed the circle of curvature at a given point on a curve. We now have general expressions for the coördinates of the center of this circle, which is called the center of curvature. The locus of centers of curvature is called the evolute. The formulas for the coördinates, calling them a and b (instead of x and y), are

$$\text{(A)}\quad a = x - dy/dx\,[1 + (dy/dx)^2]/(d^2y/dx^2),$$
$$b = y + [1 + (dy/dx)^2]/(d^2y/dx^2),$$

where x and y are the coördinates of the point on the curve. The student is reminded that he should calculate the numerical values of dy/dx and d^2y/dx^2 for the point, before substituting in (A). Observation of the rule to *evaluate numerically as soon as possible* will save much labor in many places. In certain cases, (a) where dy/dx is infinite and (b) where x is given explicitly in terms of y, it is more convenient to use the alternate forms

$$\text{(B)}\quad a = x + [1 + (dx/dy)^2]/(d^2x/dy^2),$$
$$b = y - dx/dy[1 + (dx/dy)^2]/(d^2x/dy^2).$$

Calculate a and b for the following at the points indicated, and check results by showing that (a, b) is a point on the normal to the curve at the given point (x, y). Also R must equal the distance from (a, b) to (x, y).

1. $y = x^2 - 6x + 10$, at $(3, 1)$.
2. $3y = x^3 - 3x^2 - 9x$, at $(3, -9)$.
3. $y = x^3 - 6x^2 + 11x - 6$, at $(1, 0)$.
4. $xy = 30$, at $(3, 10)$.
5. $y = e^x$, at $(0, 1)$.
6. $y = \sin x$, at $(\pi/2, 1)$.
7. $y = x^3 - 6x + 4$, at $(2, 0)$.
8. $y = (x^2 + 4)/x$, at $(2, 4)$.
9. $y = x^2 + 8/x$, at $(2, 8)$.
10. $y^2 - 4x - 4y + 4 = 0$, at $(0, 2)$.
11. $y^2 - x - 2y = 0$, at $(0, 2)$.
12. $x^2 + 2y^2 - 4x = 2$, at $(0, 1)$.
13. $2x^2 - y^2 + 3x + 1 = 0$, at $(0, 1)$.
14. $y^2 = 4(x^2 - 8x + 17)$, at $(4, 2)$.
15. $y^3 = 8x^2 - x^3$, at $(4, 4)$.
16. $x^2 + 2xy + 2y^2 = 10$, at $(2, 1)$.
17. $x^3 + xy^2 - 6y^2 = 0$, at $(3, 3)$.
18. $x^3 + y^3 - 4xy = 0$, at $(2, 2)$.

EXERCISE XLVII

Evolutes. Parametric Equations. When the equation of a curve is given parametrically, the formulas for a and b give expressions in t. These equations, $a = f(t)$ and $b = \phi(t)$, are the parametric equations of the evolute of the original curve. If a value t_1 is substituted in x and y, and the same value t_1 in a and b, it gives us a point (x, y) on the curve, and the corresponding point (a, b) on the evolute; that is, (a, b) is the center of curvature. It should be noted that the normal to the curve at (x, y) passes through (a, b), and the distance from (x, y)

to (a, b) is R. Recall the caution concerning the finding of d^2y/dx^2 for parametric equations. Find the parametric equation of the evolute in each of the following, and sketch the curve and at least part of the evolute:

1. $x = 2t,\ y = 2t^2 - 1.$
2. $x = 2t,\ y = t^3/3.$
3. $x = 3 - 2t,\ y = t^3 - 3.$
4. $x = 2t + 1,\ y = t^3/3.$
5. $x = 4t,\ y = 2/t.$
6. $x = 2t,\ y = 4 - 4t^2$
7. $x = t^3 - 2t,\ y = t^2.$
8. $x = 3t^2 - 1,\ y = 1 - 2t.$
9. $x = 4t + t^2,\ y = 2t^2$
10. $x = 2 - t,\ y = 9 - t^2.$
11. $x = 1 - t^2,\ y = 5 - t.$
12. $x = t + 2,\ y = t^2 - 2t - 15.$
13. $x = \sin t,\ y = t.$
14. $x = 13 \sin t,\ y = 5 \cos t.$
15. $x = 4(t - \sin t),$
    $y = 4(1 - \cos t).$
16. $x = 2 \cos t + \cos 2t,$
    $y = 2 \sin t + \sin 2t.$
17. $x = 3 \cos t + \cos 3t,$
    $y = 3 \sin t + \sin 3t.$
18. $x = \sec^3 t,\ y = \tan^3 t.$
19. $x = 3 \csc t,\ y = 4 \cot t.$
20. $x = \cos^4 t,\ y = \sin^4 t.$
21. $x = 2 - \cos t,\ y = t - \sin t.$
22. $x = t - \cos t,\ y = t - \sin t.$

CHAPTER VIII

PARTIAL DERIVATIVES. APPLICATIONS

EXERCISE XLVIII

Partial Derivatives. The essential thing here is that although the function u is a function of a number of variables x, y, t, etc., nevertheless, while taking the partial derivative with respect to x, $\partial u/\partial x$, all the variables except x are treated as constants, — and the differentiation is then the same as usual. In the following find the partial derivative of u with respect to each of the variables on the right-hand side. In an implicit function involving three or more variables, x, y, z, etc., we can find a partial derivative, say $\partial z/\partial x$, by regarding variables except x and z as constants during the process of differentiation.

1. $u = x^3 - 3x^2y + 3xy^2 - y^3$.

2. $u = x^3 + y^3 + z^3 - 3xyz$.

3. $z = 2x^3 - 5x^2y + 3xy^2 - 8y^2 - 7xy + 6x$.

4. $u = (ax^2 + bxy + cy^2)^3$.

5. $u = \log x^y$.

6. $u = y\cos x + x\cos y$.

7. $u = xe^y + ye^x$.

8. $u = \sin^2 x + \sin x \cos y + \cos^2 y$.

9. $u = \sin x \log y + \log x \cos y$.

10. $u = xy\sin z + xz\log y + e^x y$.

11. $u = 2xy - ye^{\sqrt{3-x}}$.

12. $u = 2e^{x^2y} + 3a^{xy^2} + 2x^3y^2$

13. $u = x^2y^3 + x^2\sin xy + y\log x$.

14. $e^z\cos x - \cos y = 0$, find $\partial z/\partial x$ and $\partial z/\partial y$.

15. $y^2/a - z^2/b = 2x$, find $\partial z/\partial x$ and $\partial z/\partial y$.

16. $xy + yz + zx = 16$, find $\partial z/\partial x$, $\partial y/\partial x$, $\partial z/\partial y$.

17. $u = x^3 - x^2y - 5xy^2 + y^3$, prove that $x\,\partial u/\partial x + y\,\partial u/\partial y = 3u$.

18. $u = x^2y + y^2z + z^2x$, prove that $\partial u/\partial x + \partial u/\partial y + \partial u/\partial z = (x+y+z)^2$.

19. $u = e^{xy} + \sin(x+y)$, prove that $x\,\partial u/\partial x - y\,\partial u/\partial y = x - y\cos(x+y)$.

20. $u = e^{(y^2+xy-x^2)/(y^2+xy)}$, prove that $x\,\partial u/\partial x + y\,\partial u/\partial x = 0$.

21. $u = \cos[(x+y)/2z]$, prove that $x\,\partial u/\partial x + y\,\partial u/\partial y + z\,\partial u/\partial z = 0$.

22. $u = (x^3 - y^3)/xy$, show that $x\,\partial u/\partial x + y\,\partial u/\partial y = u$.

23. $u = x^3/y - xy + x^2y^2/(x^2+y^2)$, show that $x\,\partial u/\partial x + y\,\partial u/\partial y = 2u$.

24. A circular cylinder of radius x and altitude y is surmounted by a hemisphere of the same radius. Show that (a) if the base remains constant, the volume increases numerically πx^2 times as fast as the altitude, and (b) if the altitude remains constant, the volume increases $2\pi x(x+y)$ times as fast as the radius of the base.

25. Given the frustum of a cone with R and r as the radii of the lower and upper bases respectively, and with altitude h. Show that if h and R remain constant, the volume increases numerically $\frac{1}{3}\pi h(2R+r)$ times as fast as the radius of the lower base.

EXERCISE XLIX

Differentiation of Implicit Functions by Means of Partial Derivatives. In Exercise XXI dy/dx for implicit functions was found. In general, however, it is easier to write down dy/dx at once from the fact that if $F(x, y) = 0$ is the equation, then $dy/dx = -(\partial F/\partial x) \div (\partial F/\partial y)$. Apply this theorem in finding dy/dx for each of the following (clear of fractions before differentiating).

1. $9x^2 + 4y^2 = 36$.

2. $x^3 - x^2y + xy^2 + y^3 = 0$.

3. $3x^2 + 2y^2 - 4x - 4 = 0$.

4. $x^3 + y^3 - 3axy = 0$.

5. $y^2 - (x+y)/(x-y) = 0$.

6. $y^2(x-2) - (x+2)^3 = 0$.

7. $x^{\frac{1}{2}} + y^{\frac{1}{2}} = a^{\frac{1}{2}}$.

8. $x^{\frac{2}{3}} + y^{\frac{2}{3}} = a^{\frac{2}{3}}$.

9. $2\sin^2 x + 3\sin^2 y = 0$.

10. $y\sin x - x\cos y = 0$.

11. $\cos(x+y) - xy = 0$.

12. $2x + 3y + 4e^{xy} = 0$.

13. $e^x \cos x - \cos y = 0$.

14. $x\cot y + y\sin x = 0$.

15. $ye^x + 2xy^3 = 0$.

16. $x^2 - (x^2+y^2)/(x^2-y^2) = 0$.

17. $e^y - x^{x+y} = 0$.

18. $a^{xy} + x\sin y = 0$.

19. $xe^y - y + 1 = 0$.

20. $x\log y + \sqrt{x^2+y^2} = 0$.

21. $\tan(x^2+y^2) + e^{x^2} + y^2 = 0$.

22. $x^y - y^x = 0$.

23. $y/x - \arctan x/y = 0$.

24. $a^{x-y} - x^y = 0$.

EXERCISE L

Successive Partial Derivatives. This process requires no explanation except the meaning of the symbol. Thus, $\partial^2u/(\partial x\partial y)$ means that u is first differentiated regarding x as the only variable on the right-hand side, and then this result is differentiated regarding y as the only variable. The student should be convinced that the order of differentiation is immaterial. In each example find all second partial derivatives, as $\partial^2u/\partial x^2$, $\partial^2u/\partial y^2$, $\partial^2u/\partial z^2$, $\partial^2u/(\partial x\partial y)$, $\partial^2u/(\partial x\partial z)$, etc., as well as any indicated relation.

1. $u = 2x^3 + 3x^2y - 3y^2$. Show that $\partial^2u/(\partial x\partial y) = \partial^2u/(\partial y\partial x)$.

2. $u = e^x \cos y - e^y \sin x$. Show that $\partial^2u/(\partial x\partial y) = \partial^2u/(\partial y\partial x)$.

3. $u = (ax^2 + by^2 + cz^2)^3$. Show that $\partial^3u/(\partial x^2\partial y) = \partial^3u/(\partial x\partial y\partial x)$
$= \partial^3u/(\partial y\partial x^2)$.

4. $u = \sin x \log y + \log x \cos y$.

5. $u = x^y$. Show that $\partial^2u/(\partial x\partial y) = \partial^2u/(\partial y\partial x)$.

6. $u = a^x + b^y + x^2y^2$. Find also $\partial^3u/(\partial x^2\partial y)$.

7. $u = x^3y^2 - 3xy^4 + 4x^2y^3$. Find also $\partial^3u/(\partial x\partial y^2)$.

8. $u = e^{xy} + ye^x + xe^y$

9. $u = y^2 - 2ye^x + 2x \log y$.

10. $u = ax^3y^2z + bxy^3z^4 + cy^5 + dx^2z^3$. Find $\partial^4u/(\partial x^2\partial y\partial z)$.

11. $u = 2x^3 - 5x^2y + 3xy^2 - 8y^2 - 7xy + 6x$. Find $\partial^3u/\partial x^3$, $\partial^3u/(\partial x^2\partial y)$, $\partial^3y/(\partial x\partial y^2)$.

12. $u = \dfrac{xy}{x+y}$. Show that $x^2\partial^2u/\partial x^2 + 2xy\partial^2u/(\partial x\partial y) + y^2\partial^2u/\partial y^2 = 0$.

13. $u = \log\sqrt{x^2 + y^2}$. Show that $\partial^2u/\partial x^2 + \partial^2u/\partial y^2 = 0$.

14. $u = 1/\sqrt{x^2 + y^2 + z^2}$. Show that $\partial^2u/\partial x^2 + \partial^2u/\partial y^2 + \partial^2u/\partial z^2 = 0$.

15. If $y = e^{ax+bt+c}$, show that if $\partial^2y/\partial x^2 = A\,\partial^2y/\partial t^2 + B\,\partial y/\partial t$, then $a^2 = Ab^2 + Bb$.

16. If $u = \log r$, where $r^2 = (x-a)^2 + (y-b)^2$, and $x - a$ and $y - b$ are not zero simultaneously, show that $\partial^2u/\partial x^2 + \partial^2u/\partial y^2 = 0$.

17. If $u = 1/r$, where $r^2 = (x-a)^2 + (y-b)^2 + (z-c)^2$, and $(x-a)$, $(y-b)$, and $(z-c)$ do not vanish simultaneously, prove the relation $\partial^2u/\partial x^2 + \partial^2u/\partial y^2 + \partial^2u/\partial z^2 = 0$.

18. Euler's theorem for homogeneous functions in two variables states that "if $u = f(x, y)$ is homogeneous, $x\,\partial u/\partial x + y\,\partial u/\partial y = nu$, where n is the total degree of u." Use this criterion to test (a) $(x^2 + y^2)/(y - x)$, (b) $e^{xy/(x^2+y^2)}$, (c) $\sin(xy)$, for homogeneity.

19. Show that $y = \phi(x + \kappa t) + \psi(x - \kappa t)$ satisfies the functional equation $\partial^2 y/\partial t^2 = k^2\,\partial^2 y/\partial x^2$ for all forms of ϕ and ψ.

EXERCISE LI

Total Differential and Total Derivative. The total differential is defined thus, where u is a function of any number of variables x, y, z, etc:

$$(1)\quad du = \partial u/\partial x \cdot dx + \partial u/\partial y \cdot dy + \partial u/\partial z \cdot dz + \cdots.$$

The total derivative, similarly, with respect to any of the variables is

$$(2)\quad du/dx = \partial u/\partial x \cdot dx/dx + \partial u/\partial y \cdot dy/dx + \partial u/\partial z \cdot dz/dx + \cdots.$$

Or the differentiation may be with respect to some other independent variable, as t, that is,

$$(3)\quad du/dt = \partial u/\partial x \cdot dx/dt + \partial u/\partial y \cdot dy/dt + \partial u/\partial z \cdot dz/dt + \cdots.$$

Write down the total differential for the following:

1. $u = x^2y - 2xy - y^2 - x$.

2. $u = \sin(x + y) + \cos(x - y)$.

3. $u = y \sin x + \cos(x - y)$.

4. $u = e^{xyz}$.

5. $u = \sqrt{x^2 + y^2} + \sin x \sin y$.

6. $u = x^3 - 3x^2y + y^3$.

7. $u = x^2y/(4 - z^2)$.

8. $u = x^2y - xz^2 + z^2y - xyz$.

9. $u = (1/\sqrt{x - 1}) + yz$.

10. $u = \text{arc tan}[(2x + y - x^2y)/(1 - 2xy - x^2)]$.

11. $u = a^x b^y + \sin xy$.

12. $u = \text{arc cos}[(1 - xy)/\sqrt{1 + x^2 + y^2 + x^2y^2}]$.

13. $u = \text{arc cos}\, z/xy$.

14. $u = \text{arc tan}\, xy + x \sin y + x^y$.

In the next six examples write down the total derivative with respect to both x and t.

15. $u = x^2 - 2xy + y^2 + 4$.

16. $u = 3x^3 - xy^2 + x^2z + y^2z$.

17. $u = e^{-x} \sin y + \sin x$.

18. $u = \sin x \cos y + xy$.

19. $u = \log(x^2 + 2xy - y^2)$.

20. $u = \text{arc tan}\, xy + xy$.

Sometimes, in addition to the function, we know the relation connecting the variables on the right-hand side, in which case the total derivative may be written in terms of one variable (do not, however, substitute before differentiating the original expression by aid of partial derivatives). In other cases x, y, z, etc. may be expressed in terms of some extraneous variable. Find du/dt or du/dx in each of the following:

21. $u = x^2 - t^2$, $x = \tan t$. **22.** $u = 2^{ty}$, $y = \log t$.

23. $u = 2x^2 + 3y^2 + xy$, $x = \tan 2t$, $y = \sin^2 t$.

24. $u = xy + yz + xz$, $y = e^x$, $z = a^x$.

25. $u = 2^x + 3^y + 4^z$, $x = \sin t$, $y = \cos t$, $z = \tan t$.

EXERCISE LII

Total Derivatives and Differentials. Application to Rates and Small Errors. The idea of total derivatives is directly applicable to rate problems. For example, let $F(x, y, z) = 0$ be the known relation connecting the variables x, y, z. Then $(\partial F/\partial x)(dx/dt) + (\partial F/\partial y)(dy/dt) + (\partial F/\partial z)(dz/dt) = 0$, where dx/dt, dy/dt, dz/dt are the rates at which the variables x, y, z, respectively, vary relative to t. Or take the case where one variable function, such as cost, volume, area, etc., depends on a number of variables, such as $A = F(x, y, z)$. Here the total derivative dA/dt is given by $dA/dt = (\partial F/\partial x)(dx/dt) + (\partial F/\partial y)(dy/dt) + (\partial F/\partial z)(dz/dt)$, and dA/dt will be the rate at which the function A changes. In time-rate problems the independent variable t is the time. Apply to some of the examples in Exercise XL.

Total differentials are used similarly in calculating approximately small differences or errors in a function, due to slight variations or inaccuracies in the independent variables. For example, the area of a rectangle is given by $A = xy$, and therefore (1) $dA = y\,dx + x\,dy$, where x and y are the measurements and dx and dy the errors or inaccuracies in measuring. dA is, then, the approximate error in area due to the errors dx and dy.

As a numerical example, suppose the measured dimensions are 3×4 in., with a maximum error of .01 in. in measuring. Then $dx = dy = \pm .01$. Therefore $dA = .04 + .03 = .07$ sq. in. is approximately the maximum error in area. This method is applicable to any number of variables, the general form being

$$(1)\quad dA = (\partial F/\partial x)dx + (\partial F/\partial y)dy + (\partial F/\partial z)dz + \cdots,$$

where $A = F(x, y, z \cdots)$. The first term may be called the error in A due to the error in x, the second term that due to the error in y, etc. The maximum error in A is the sum of the absolute values of the terms on the right in (1).

The value dA given by (1) is the **absolute error.** The **proportional, or relative, error** is dA/A. Finally, the **percentage error** is $100(dA/A)$. The student should observe that the proportional error is the logarithmic derivative, that is, the derivative of $\log A$, and use this fact in his work.

1. A point on the plane $x + 2y + 3z - 5 = 0$ moves parallel to the XOZ plane. When $x = 4$ and is increasing 5 units per second, find (a) the rate at which z is changing and (b) the speed at which the point is moving and its direction.

2. A point is moving on the intersection of the sphere $x^2 + y^2 + z^2 = 49$ and the plane $y = 2$. When $x = 6$ and is increasing 4 units per second, find (a) the rate at which z is changing and (b) the speed at which the point is moving.

3. A point is moving on the surface $z = x^{\frac{1}{2}} + y^{\frac{1}{2}}$ and also on the plane $y = 4$. When $z = 3$ and is increasing 2 units per second, find the rate at which x is changing and the speed at which the point moves.

4. A particle is moving on the intersection of the surface $x^2 + xy + y^2 - z^2 = 0$ and the plane $x - y + 2 = 0$. At the time when $x = 3$ and is increasing 2 units per second, at what rate are y and z changing? What is the total speed of the particle?

5. A particle is moving on the intersection of the surface $xyz = 8$ and the plane $2x - y = 0$. If at a given time $z = 4$ and is increasing 2 units per second, how fast are x and y changing? What is the speed?

6. What is the approximate error in the volume and surface of a cube of edge 6 in. if an error of .02 in. is made in measuring the edge?

7. The formulas for the surface and volume of a sphere are $S = 4\pi r^2$ and $V = \frac{4}{3}\pi r^3$. If the radius is found to be 3 in. by measuring, (a) what is the maximum error in S and V if measurements are accurate to .01 in.? (b) What is the maximum percentage error in each case?

8. The total surface of a cylinder with diameter equal to altitude is to be gilded at a cost of 10¢ per square inch. If the altitude is measured as 24 in., what is the maximum error in cost, measurement being accurate to 1/32 in.?

9. A particle is moving on the ellipse $x^2 + 4y^2 = 20$ and is at the point (2, 2). What is the approximate change in y due to a change of 0.1 in x?

10. A particle is moving on the surface $x^2 + 2y^2 - 4z^2 + 8z + 10 = 0$ and is at the point (2, 3, 4). If the x coördinate is increased 0.1 and the y coördinate is decreased 0.1, what is the approximate change in z?

11. In measuring a box I find the measurements to be $3 \times 4 \times 5\frac{1}{2}$ in. If my measurements are accurate to within .01 in., (a) what is the maximum error in volume? (b) What is the maximum percentage error?

12. Given the surface $z = xy/(x + y)$. If, at the point where $x = y = 4$, x and y are each increased by 1/10, what is the approximate change in z?

13. Given the surface $x^2 - 2y^2 - 2z = 0$ and the point (2, 1, 1) on it. What is the c ige in z if x is increased .01 and y is decreased by .02?

14. The specific gravity of a solid is given by the formula $s = P/w$, where P is the weight in a vacuum and w is the weight of an equal volume of water. How is the specific gravity affected by an error of $\pm 1/10$ in weighing P and $\pm 1/20$ in weighing w, assuming $P = 8$ and $w = 1$ in the experiment, (a) if both errors are positive? (b) if one error is negative? (c) What is the maximum percentage error?

15. If specific gravity is determined by the formula $s = A/(A - W)$, where A is the weight in air and W the weight in water, what is (a) approximately the maximum error in s if A can be read within .01 lb. and W to .02 lb., the actual readings being $A = 9$ lb., $W = 5$ lb.? (b) the maximum relative error?

16. The resistance of a circuit was found by using the formula $C = E/R$, where C = current and E = electromotive force. If there is an error of 1/10 ampere in reading C and 1/20 volt in reading E, what is the error in R if readings are $C = 15$ amperes and $E = 110$ volts? What is the maximum percentage error?

17. If the formula $\sin(x + y) = \sin x \cos y + \cos x \sin y$ were used to calculate $\sin(x + y)$, what approximate error would result if an error of 0.1 were made in measuring both x and y, the measurements of the two acute angles giving $\sin x = 3/5$ and $\sin y = 5/13$?

18. The acceleration of a particle down an inclined plane is given by $f = g \sin \alpha$. If g varies by $\frac{1}{10}$ ft., and α, which is measured as 30°, may be in error .01°, what is the maximum error in f? Take the normal $g = 32$ ft./sec².

19. The period of a pendulum is $P = 2\pi\sqrt{L/g}$. What is the greatest error in the period if there is an error of $\pm 1/10$ ft. in measuring a 10-foot suspension, and g, taken as 32 ft./sec.², may be out 1/20 ft. per sec.²? What is the percentage error?

20. The length L and the period P are connected by the equation $4\pi^2 L = P^2 g$. If L is calculated assuming $P = 1$, and $g = 32$ ft./sec.², what is the approximate error in L if the true values are $P = 1.02$ and $g = 32.01$ ft./sec.²? What is the percentage error?

21. A triangle ABC is being transformed so that the angle A changes from 0° to 90° in 10 sec., while side AC decreases 1 in./sec. and side AB increases 1 in./sec. If at the time of observation $A = 60°$, $AC = 16$ in., and $AB = 10$ in., (a) how fast is BC changing? (b) How fast is the area of ABC changing?

22. A crank OA and a connecting rod AB move a slider, attached by a pivot at B, in a straight path in the direction of O, the center of the crank shaft. (a) How fast is the slider B moving when the crank and connecting rod are at right angles, assuming the crank to be rotating with constant angular velocity? (b) Evaluate the result in (a), assuming a 5-foot crank, a 13-foot rod, and an angular velocity of 100 R.P.M.

23. Assume in example 22 that the machine is eccentric; that is, the path L of the slider is not directed through O. Take the perpendicular distance from O to L as c, the length of the crank as a, of the rod as b, and the constant angular velocity as ω. (a) Derive a general expression for v, the velocity of the slider B, assuming that the crank is horizontal at the start. (b) What is the value of this velocity at the start, and also when A is at its highest point? (c) Calculate the angular velocity of the rod about B at each of these times. Evaluate for some particular values of the constants.

Sometimes the errors in quantities are not errors of observation, but are due to defects in the instruments, such as tapes, chains, etc. Here the original errors are usually percentage or proportional ones, as in the examples immediately following.

24. The dimensions of a cone are radius 4 in. and altitude 6 in. What is the error in volume and total surface if there is a shortage of .01 in. per inch in the measure used?

25. A solid is in the form of a cylinder capped at each end with a hemisphere of the same radius as the cylinder. Its measured dimensions are diameter 6 in. and total length 16 in. What is the approximate error in volume and surface if the tape used in measuring had stretched uniformly $\frac{1}{2}\%$ beyond its proper length?

26. A dealer, in estimating the number of bricks in a pile, measures it to be $8 \times 50 \times 5$ ft. counting 12 bricks to the cubic foot. If he actually stretched the tape 2% beyond normal in measuring, how much did he profit, bricks being worth $10 per thousand?

27. A solid as measured consists of a right cone with radius of base 6 in. and altitude 12 in., from which has been hollowed out a coaxial cone with radius 4 in. and altitude 9 in. If the tape used in measuring is so elastic that there may be an error of $\pm \frac{1}{20}$ in. per inch, what is the maximum error in calculating the volume of the solid? the maximum relative error?

28. It is found that a pendulum 99.245 cm. long has a period of exactly 2 sec. when $g = 980$ cm./sec.^{2} If an error of 0.3 cm. is made in measuring l, and g is 979.2, what will the pendulum gain or lose in 10,000 beats?

29. In calculating the length of a line the formula $s^2 = (x_1 - x_2)^2 + (y_1 - y_2)^2$ is used. (a) Calculate the percentage error if the following data are given: $s = 3000$ ft.; errors in x_1, x_2, y_1, y_2 are respectively 1 ft., -2 ft., 2 ft., and 3 ft.; and the inclination of the line is 36°. (b) What is the maximum error and the percentage error if the errors are ± 1 ft., ± 2 ft., ± 2 ft., and ± 3 ft. respectively?

30. In measuring the strength J of a galvanic current by means of a tangent galvanometer the formula is $J = c \tan \alpha$, where c is a known constant of the instrument and α is the deflection of the needle and therefore subject to errors of observation. (a) What is the general expression for the error in J due to a small error in α? (b) Assuming that α can be read with equal accuracy on all parts of the scale, at what readings would the absolute error in J be a minimum? (c) Where would the proportional or percentage error be a minimum?

31. A Bunsen grease-spot photometer is often used to compare the intensities of two sources of light. The two light sources of intensities I_1 and I_2 are placed at a fixed distance a apart, and a vertical paper screen with a well-defined grease spot (usually near its center) can be moved back and forth on a slide until the grease spot seems to be equally illumined on both sides and apparently disappears. If a_1 and a_2 are the distances from the screen to the two light sources $(a_1 + a_2 = a)$, the formula used to obtain I_2 when I_1 is a standard of known intensity is $I_1/I_2 = a_1^2/a_2^2$. (a) Get an expression for the error in the value of I_2

due to a small error in measuring a_2. (b) In what position of the screen would the absolute error in I_2 be a minimum? (c) Where is the percentage error a minimum? (d) Calculate the absolute and the percentage error if the measurements are $I_1 = 16$ candle power, $a = 10$ meters, $a_2 = 746$ cm. with a possible error of ± 1 cm. (e) Get an expression for the error in I_2 if there is an error in I_1. (f) What is the maximum error and the maximum percentage error in I_2, using the data of (d) and assuming also a possible error of $\pm \frac{1}{4}$ candle power in I_1.

32. Given the gas formula $pV = k\theta$, where p is pressure, V is volume, θ is temperature, and k is a constant. (a) If the volume is kept constant, how does the pressure change per degree of change in temperature? (b) If the pressure is kept constant, how does the volume change per degree of increase in temperature? (c) If at a given time $\theta = 250°$ and $V = 100$ liters, what is the proportional change in P if θ increases 5° and V decreases 1 liter? (d) If under the conditions of (c) the temperature increases 10° per minute and the volume is kept constant, what is the percentage rate of increase of the pressure?

33. The quantity of heat liberated when x molecules of sulphuric acid are mixed with y molecules of water is given by $Q = ay/(bx + y)$. (a) What is the increase in Q per added molecule of water if the amount of acid is kept constant? (b) What is the increase in Q per added molecule of acid if the amount of water is kept constant? If at a given time there are 10 times as many molecules of acid as of water, what is the proportional change in Q if the amount of acid is increased 10% and the amount of water 5%?

EXERCISE LIII

Exact Differentials. Where an expression $Mdx + Ndy$ is the total differential of some function of x and y, it is called an *exact differential.* If $f(x, y)$ is the function, M and N are $\partial f/\partial x$ and $\partial f/\partial y$. Hence it is easily seen, since $\partial^2 f/(\partial x\,\partial y) = \partial^2 f/(\partial y\,\partial x)$, that the condition for $Mdx + Ndy$, to be exact, is

$$\partial M/\partial y = \partial N/\partial x.$$

This fact is of importance later in finding the function when the exact differential is given. Determine which of the following are exact differentials:

1. $(1 + x^2)\,y\,dy + (y^2 - 3)\,x\,dx$. **2.** $2 \arctan y/x \cdot dx + \log(x^2 + y^2)\,dy$.

3. $y \sin 2x\,dx + \sin^2 x\,dy$.

4. $ye^x[\sin(x + y) + \cos(x + y)] \cdot dx + e^x[y \cos(x + y) + \sin(x + y)]dy$.

5. $(2xy \sin x + x^2y \cos x)\,dx + x^2 \sin x\,dy$.

6. $\sin 2x \cos^2 y\,dx + \sin^2 x \sin 2y\,dy$.

7. $(x^2 - 4xy - 2y^2)\,dx + (y^2 - 4xy - 2x^2)\,dy$.

8. $(y - x)/y^2 \cdot dy + 1/y \cdot dx$.

9. $(ye^x + e^y)\,dx + (e^x + xe^y)\,dy$.

10. $\left(y + \sqrt{x^2 - y^2}\right)/\left(x\sqrt{x^2 - y^2}\right) \cdot dx - 1/\sqrt{x^2 - y^2} \cdot dy$.

11. $(x^3 + 3x^2y^2 - y^2)\,dx + (x^2 + 3x^2y - 2xy)\,dy$

12. $(y^3 - 2xy)\,dx + (3xy^2 - x^2)\,dy$.

13. $(x^3 + xy^2 - y)/(x^2 + y^2) \cdot dx + (x^2y + x + y^3)/(x^2 + y^2) \cdot dy$.

14. $\text{arc}\sin(x/y)\,dx + \sqrt{y^2 - x^2}/y \cdot dy$.

15. $(x^3 - 3xy^2)\,dx + (3x^2y - y^3)\,dy$.

16. $(1 - e^{y/x}/x)\,dy + ye^{y/x}/x^2 \cdot dx$.

17. $(6x - 2y + 1)\,dx + (2y - 2x - 3)\,dy$.

18. $(2xy + 1)/y \cdot dx + (y - x)/y^2\,dy$

19. $(e^{y^2} - 2x \sin y)\,dx + (2xye^{y^2} - x^2 \cos y)\,dy$.

20. $(y/x - 2xy + y \sin x)\,dx + (\log x - x^2 - \cos x)\,dy$

CHAPTER IX

SERIES

EXERCISE LIV

Convergence of Number Series. There are two common tests for convergence, given by these rules: a series is convergent

(a) if it is less, term for term, than a known convergent series;

(b) if the limit of the ratio of the $(n+1)$th term and the nth term of the series is *definitely* less than unity as n increases indefinitely.

The first (a) is known as the comparison test, and the second (b) as Cauchy's ratio test. Correspondingly a series is divergent

(a) if it is greater, term for term, than a known divergent series;

(b) if the limit of the ratio of the $(n+1)$th term and the nth term of the series is *definitely* greater than unity as n increases indefinitely.

In using Cauchy's ratio test we find that the case where the limiting value of the ratio of the $(n+1)$th and nth terms is exactly 1 is left undetermined, and must be settled by the comparison, or some other, test.

When the law of the series is not given there remains the practical difficulty of determining the nth term. Note that the nth and $(n+1)$th terms mean two successive general terms, as any finite number of terms may be lopped off at the beginning of a series without affecting its convergence or nonconvergence. This fact should be remembered in comparing two series. In the case of an alternating series, it is sufficient for convergence to know that the terms of the series are decreasing.

The following are the more common series used in the comparison test.

Convergent:

(a) Any geometrical progression with ratio less than unity, as $1 + 1/2 + 1/4 + 1/8 + \cdots$, where $r = 1/2$.

(b) The p series where p is greater than 1,

$$1/1^p + 1/2^p + 1/3^p + 1/4^p + \cdots 1/n^p + \cdots.$$

Divergent:

(c) The harmonic series,

$$1 + 1/2 + 1/3 + 1/4 + 1/5 + 1/6 + \cdots.$$

(d) The p series where p is less than 1. Note that (c) is really the case where $p = 1$.

Test the following series:

1. $1/3 + 2/3^2 + 3/3^3 + \cdots n/3^n + (n+1)/3^{n+1} + \cdots.$

2. $1/1 + 1/\sqrt{2} + 1/\sqrt{3} + \cdots 1/n^{\frac{1}{2}} + 1/(n+1)^{\frac{1}{2}} + \cdots.$

3. $1/1 + 1/\sqrt{2^3} + 1/\sqrt{3^3} + \cdots 1/n^{\frac{3}{2}} + 1/(n+1)^{\frac{3}{2}} + \cdots.$

4. $2/10 + 3/10^2 + 4/10^3 + \cdots n/10^n + \cdots.$

5. $1 + 1/2^2 + 1/3^3 + 1/4^4 + \cdots 1/n^n + \cdots.$

6. $1/(1\cdot 2) + 1/(2\cdot 3) + 1/(3\cdot 4) + 1/(4\cdot 5) + \cdots 1/n(n+1) + \cdots.$

7. $2/(2\cdot 3) + 4/(3\cdot 4) + 6/(4\cdot 5) + \cdots 2n/(n+1)(n+2) + \cdots.$

8. $2/(2\cdot 3\cdot 4) + 4/(3\cdot 4\cdot 5) + 6/(4\cdot 5\cdot 6)$
$+ \cdots 2n/(n+1)(n+2)(n+3) + \cdots.$

9. $1/3 + 1/6 + 1/9 + \cdots 1/3n + \cdots.$

10. $1/3 + (1\cdot 2)/(3\cdot 5) + (1\cdot 2\cdot 3)/(3\cdot 5\cdot 7)$
$+ \cdots n!/[3\cdot 5\cdot 7\cdots(2n+1)] + \cdots.$

11. $1/1 + (1\cdot 3)/(1\cdot 4) + (1\cdot 3\cdot 5)/(1\cdot 4\cdot 7)$
$+ \cdots [1\cdot 3\cdot 5\cdots(2n-1)]/[1\cdot 4\cdot 7\cdots(3n-2)] + \cdots.$

12. $1/1 + (1\cdot 2)/(1\cdot 3) + (1\cdot 2\cdot 3)/(1\cdot 3\cdot 5)$
$+ (1\cdot 2\cdot 3\cdot 4)/(1\cdot 3\cdot 5\cdot 7) + \cdots.$

13. $2/3 + 2(2/3)^2 + 3(2/3)^3 + 4(2/3)^4 + \cdots.$

14. $1/5 + 1/(5+1) + 1/(5+2) + 1/(5+3) + \cdots 1/(5+n)\cdots.$

15. $1/\sqrt{3} + 1/\sqrt[3]{3} + 1/\sqrt[4]{3} + \cdots 1/\sqrt[n]{3} + \cdots.$

16. $2/4 + (2\cdot 4)/(4\cdot 7) + (2\cdot 4\cdot 6)/(4\cdot 7\cdot 10)$
$+ \cdots 2n!/[4\cdot 7\cdot 10\cdots(3n+1)]\ \cdots.$

17. $1/(2+1)+1/(2^2+1)+1/(2^3+1)+\cdots 1/(2^n+1)+\cdots$.

18. $1/(1^2+1)+1/(2^2+1)+1/(3^2+1)+\cdots 1/(n^2+1)+\cdots$.

19. $1/(3-2)+1/(3^2-2)+1/(3^3-2)+\cdots 1/(3^n-2)+\cdots$.

20. $3/(2\cdot 3)+6/(3\cdot 4)+9/(4\cdot 5)+12/(5\cdot 6)$
$+\cdots 3n/[(n+1)(n+2)]+\cdots$.

21. $1+1/2+1/6+1/12+1/20+1/30+1/42+\cdots$.

22. $1+5/2!+9/3!+13/4!+17/5!+\cdots$.

23. $2/7+2^2/26+2^3/63+2^4/124+\cdots$.

24. $1/(1\cdot 2\cdot 3\cdot 4)+1/(2\cdot 3\cdot 4\cdot 5)+1/(3\cdot 4\cdot 5\cdot 6)+\cdots$.

EXERCISE LV

Interval or Region of Convergence. Power Series. In determining the values of x for which a power series converges, it is necessary to know two consecutive general terms, call them A_n and A_{n+1}, both of which will involve n and x. Then the ratio of these terms is given by $R = A_{n+1}/A_n$, and for convergence the limiting value of R, as n increases indefinitely, must be definitely less than unity in numerical value. Therefore find the limiting value of A_{n+1}/A_n, as n approaches infinity, x being regarded as a *constant*. This limiting value of R will in general be a function of x. Call this $R(x)$. The power series will then converge for all values of x which make $R(x)$ less than 1 in absolute value. The end points, that is, the values of x which make $R(x)=1$, must be tested separately, as in the preceding exercise. If $R(x)\doteq 0$, the series converges for all values of x; if $R(x)\doteq\infty$, it converges for no values of x (except $x=0$).

Determine the interval of convergence for the following series:

1. $1-2x+3x^2-4x^3+\cdots$.

2. $1+x+x^2/2+x^3/3+\cdots$.

3. $1+x+x^2/2^2+x^3/3^2+x^4/4^2+\cdots$.

4. $2x+3x^2/2!+4x^3/3!+\cdots(n+1)x^n/n!+\cdots$.

5. $x/(1\cdot 3)+x^2/(2\cdot 3^2)+x^3/(3\cdot 3^3)+x^4/(4\cdot 3^4)+\cdots x^n/(n\cdot 3^n)+\cdots$.

6. $1+x^2/2+[(1\cdot 3)/(2\cdot 4)]x^4+[(1\cdot 3\cdot 5)/(2\cdot 4\cdot 6)]x^6+\cdots$.

7. $x+x^4+x^9+\cdots x^{n^2}+\cdots$

8. $1+x+x^2/2^2+x^3/3^3+x^4/4^4+\cdots$.

9. $2x/2 + 2^2x^2/5 + 2^3x^3/10 + 2^4x^4/17 + \cdots 2^nx^n/(n^2+1) + \cdots.$

10. $1 + x + x^2/2! + x^3/3! + x^4/4! + \cdots.$

11. $1 + 4x + 9x^2 + 16x^3 + \cdots.$ **12.** $x - x^3 + x^5 - x^7 + \cdots.$

13. $1 - 3x^2 + 5x^4 - 7x^6 + \cdots (-1)^{n-1}(2n-1)^{2(n-1)} + \cdots.$

14. $1 + x/(2 \cdot 1) + x^2/(2^2 \cdot 2) + x^3/(2^3 \cdot 3) + x^4/(2^4 \cdot 4) + \cdots.$

15. $1/3 + 2x/(2 \cdot 3^2) + 3x^2/(2^2 \cdot 3^3) + 4x^3/(2^3 \cdot 3^4) + \cdots.$

16. $2x^2/3 + 4x^3/8 + 8x^4/15 + 16x^5/24 + \cdots 2^{n-1}x^n/n^2 - 1 \cdots.$

17. $x/3^{\frac{1}{2}} + x/3^{\frac{1}{3}} + x^2/3^{\frac{1}{4}} + x^3/3^{\frac{1}{5}} + \cdots.$

18. $x/(1 \cdot 2) + 2x^2/(2 \cdot 2 \cdot 3) + 3x^3/(2^2 \cdot 3 \cdot 4) + 4x^4/(2^3 \cdot 4 \cdot 5) + \cdots.$

The binomial theorem gives rise to an infinite power series in x when the exponent n is not an integer. We will use this theorem in the form

$$(1)\quad (1+x)^n = 1 + nx + \frac{n(n-1)}{2!}x^2 + \frac{n(n-1)(n-2)}{3!}x^3 + \frac{n(n-1)(n-2)(n-3)}{4!}x^4 + \cdots.$$

Clearly any binomial expression of the form $(a + bx)^n$ can be expanded by (1), by rewriting it in the form $a^n(1 + bx/a)^n$. This leads to a method of extracting higher roots without recourse to a table of logarithms.

Expand each of the following, and determine the interval of convergence.

19. $(8 + x)^{\frac{1}{3}}$. This equals $2(1 + x/8)^{\frac{1}{3}}$.

20. $(4 + x)^{\frac{1}{2}}$. **21.** $(9 + x)^{\frac{1}{2}}$.

22. $(9 + 4x)^{\frac{1}{2}}$. **23.** $(16 - x)^{\frac{1}{4}}$.

24. $(8 + x)^{\frac{2}{3}}$. **25.** $(16 + x)^{\frac{3}{4}}$.

EXERCISE LVI

Expansion in Powers of $(x - a)$. Taylor's Theorem. It is sometimes very desirable to have a function expanded in powers of $(x - a)$, where a is some constant. This is accomplished by Taylor's theorem,

$$F(x) = F(a) + F'(a)(x-a) + \frac{F''(a)}{2!}(x-a)^2 + \frac{F'''(a)}{3!}(x-a)^3 + \cdots.$$

That is, the successive coefficients are simply the successive derivatives of $F(x)$, evaluated for $x = a$. The series may be finite or infinite. Observe that a may be a positive or negative quantity. Expand the following in terms of the quantities indicated.

1. $\tan x$ in powers of $(x-\pi/4)$; of $(x+\pi/4)$; of $(x+2)$.
2. $\sin x$ in powers of $(x-\pi/3)$; of $(x+\pi/6)$; of $(x-\pi/4)$.
3. $x^4-3x^3-x^2-5x-11$ in powers of $(x-2)$.
4. x^4-5x^2+8x-7 in powers of $(x+1)$; of $(x-2)$.
5. x^5-2x^3+4x in powers of $(x+1)$.
6. e^x in powers of $(x+3)$.
7. e^{3x} in powers of $(x-4)$.
8. $\log(1+x)$ in powers of $(x-1)$.
9. $\text{arc}\tan x$ in powers of $(x-3)$.
10. $\sin^2 x$ in powers of $(x+1)$.
11. $\cos 3x$ in powers of $(x-\pi/6)$.
12. e^{-2x} in powers of $(x-3)$.
13. $\frac{1}{2}(e^x+e^{-x})$ in powers of $(x-1)$.
14. $1/(1-x)$ in powers of $(1+x)$.
15. $x/(1+x^2)$ in powers of $(x+2)$.
16. $(x+1)/(x-2)$ in powers of $(x-3)$.
17. $\sin(x+1)$ in powers of $(x+2)$.
18. e^{x^2} in powers of $(x+1)$.
19. $\sin 2x$ in powers of $(x-\pi/6)$; of $(x-\pi/4)$.
20. $\cos x/2$ in powers of $(x+\pi/3)$; of $(x-\pi/3)$.

EXERCISE LVII

Expansion in a Power Series in *x*. This theorem, known as Maclaurin's theorem, is really a particular case of Taylor's, where $a=0$, the formula becoming

$$F(x)=F(0)+F'(0)x+F''(0)x^2/2!+F'''(0)x^3/3!+\cdots$$

A few hints: If a product consists of a power of x (or even a simple polynomial in x) and a transcendental function, expand the transcendental part and multiply by the polynomial. In some other products — since in general we use only a few terms — it is simpler to expand each term of the product separately and then multiply. A careful inspection of the quantity to be expanded should determine the method.

The following also contains a valuable hint. Suppose we expand $\log\cos x$ as a power series in x. Now differentiate both sides and you have $\tan x =$ a power series in x without further trouble. Differentiate both sides again and you have $\sec^2 x =$ a power series in x.

If the original function F is a function of $(x + h)$, the above theorem can be used to expand it as a power series in x or in h, according as x or h is regarded as the variable. The form so frequently given, $f(a + x) = f(a) + f'(a)x + f''(a)x^2/2! + \cdots$, is a useless burden to the memory as it is covered by Maclaurin's series, since the $\sin(a + x)$ can just as well be regarded as a function of x as of $(a + x)$. In fact, a *thorough* understanding of Taylor's formula as given in the preceding exercise is all-sufficient for any expansion. Some may prefer to use the method of undetermined coefficients.

There is no need to use this method, this is, Maclaurin's, if the binomial expansion is available. It is well to check up on a few such examples (as 8) by using both methods. *If at any stage the derivative can be expanded by the binomial theorem, it should be done*, and subsequent derivatives taken from this expanded form. See Exercise LV for the binomial theorem.

Expand the following:

1. $\sin 2x$.
2. e^{3x}.
3. $\cos(x/2 + \pi/6)$.
4. $\sin(x - \pi/3)$.
5. $\sin(x + \pi/3)$.
6. xe^{-x}.
7. $\log(1 - x)$.
8. $\sqrt[3]{1 + x}$.
9. $\log \cos x$.
10. $\text{arc} \tan x$.
11. $\sec^2 x$.
12. $(1 + x)\log(1 + x)$.
13. $\sqrt{1 + x^3}$.
14. $\sqrt{1 - 2x^2}$.
15. $1/\sqrt{1 - x}$.
16. $\frac{1}{2}(e_x + e^{-x})$.
17. $\frac{1}{2}(e^x - e^{-x})$.
18. $\log(x + \sqrt{1 + x^2})$.
19. $(1 - \cos x)/x^2$.
20. $(1 + x)e^x$.
21. $e^x \cos x$.
22. $e^x \log(1 + x)$.
23. $e^x/(1 + x)$.
24. $\text{arc} \tan 1/x$.

EXERCISE LVIII

Calculations by Series. Expansion into series is, in general, useful in calculations only when the series is convergent. In many instances more than one series is available. For example, in calculating $\sin 46°$, we may (a) expand $\sin(\pi/4 + x)$ as a power series in x and then put $x = 0.01745$, or (b) expand $\sin x$

in powers of $(x - \pi/4)$ and then put $x = 46(0.01745)$. In general, (a) is preferable. The following series are most frequently used and should be memorized:

$$\left.\begin{aligned}\sin x &= x - x^3/3! + x^5/5! - x^7/7! + \cdots,\\ \cos x &= 1 - x^2/2! + x^4/4! - x^6/6! + \cdots,\\ e^x &= 1 + x + x^2/2! + x^3/3! + x^4/4! + \cdots,\end{aligned}\right\}\text{ convergent for all values of } x.$$

$$\log(1 + x) = x - x^2/2 + x^3/3 - x^4/4 + x^5/5 - \cdots,$$

$$\text{convergent for } -1 < x \leq 1.$$

$$(a + bx)^n = a\ (1 + bx/a)^n = a^n[1 + (nb/a)x + \cdots].$$

$$\log\frac{M}{N} = 2\left[\left(\frac{M-N}{M+N}\right) + \frac{1}{3}\left(\frac{M-N}{M+N}\right)^3 + \frac{1}{5}\left(\frac{M-N}{M+N}\right)^5 + \cdots\right].$$

This last series can be put in the form below for use in calculating logarithms:

$$\log M = \log N + 2\left(x + \frac{x^3}{3} + \frac{x^5}{5} + \frac{x^7}{7} + \cdots\right),$$

where $x = (M - N)/(M + N)$, and it should be remembered in this form. The smaller $M - N$ is, the more rapidly the series converges.

In the following problems master the principle underlying any simplification or method suggested, and do not employ it blindly. In calculations giving rise to alternating series it is well to know that the error due to taking but n terms of the series is less than the nth term.

Calculate the following by the use of series:

1. $\sin 46°$ (both series).
2. $\cos 44°$.
3. $\sin 33°$.
4. log of all numbers up to 15.
5. $\sin 1.5°$.
6. $e^{\frac{1}{2}}$.
7. $e^{0.3}$.
8. $\log_e 1.2$.
9. $\sqrt[3]{1.2}$ by using $\sqrt[3]{1+x}$ series.
10. $\tan 31°$.
11. $\sin 58°$ (two different series).
12. $\sqrt[4]{16.2}$ by using the $(16 + x)^{\frac{1}{4}}$ series.
13. $\sin 29°$.
14. arc tan 2.
15. $\sqrt[5]{1.3}$ by using the $(1 + x)^{\frac{1}{5}}$ series.
16. $\sqrt[3]{8.5}$ by using the $(8 + x)^{\frac{1}{3}}$ series $= 2(1 + x/8)^{\frac{1}{3}}$.

17. $\sqrt[4]{18}$ by using the $(16+x)^{\frac{1}{4}}$ series $= 2(1+x/16)^{\frac{1}{4}}$.

18. $\sqrt[3]{30}$ by using the $(27+x)^{\frac{1}{3}}$ series $= 3(1+x/27)^{\frac{1}{3}}$.

19. The difference between an arc and its chord in a circle of radius r is given by the formula $2r(x/2 - \sin x/2)$. Expand this as a power series and calculate the difference in length when $x = 5°$ and $r = 10$ in. Result to six significant figures.

20. Calculate π by expanding $\arcsin x$ and then putting $x = \frac{1}{2}$ on both sides.

21. Calculate π by expanding $\arctan x$ and putting $x = 1$.

22. Calculate π by aid of the $\arctan x$ series, using the fact that $\pi/4 = \arctan\frac{1}{2} + \arctan\frac{1}{3}$.

23. Calculate π by aid of the $\arctan x$ series, using the fact that $\pi/4 = \arctan\frac{1}{2} + \arctan\frac{1}{4} + \arctan\frac{1}{8}$.

24. Expand $\log_{10}\sin(\pi/6 + x)$ as a power series in x, and calculate $\log_{10}\sin 31°$ to four decimal places. Given $\log_{10}\sin\pi/6 = 9.69897 - 10$ and $\log_{10}e = 0.43429$.

25. Expand $\log\tan(\pi/4 + x)$ and calculate the increase in the log from 45° to 50°, and check by a seven-place table.

26. Given $\log_e 5 = 1.6094$, calculate $\log_e 27$.

27. Given $\log_e 3 = 1.0986$ and $\log_e 5 = 1.6094$, calculate $\log_e 19$.

28. Calculate $\sqrt[5]{250}$.

29. Expand $1/(a+x)$ as a power series in a, and from it calculate the reciprocal of 101 and 96 to six decimal places.

30. Given $f(x) = x^3 - 7x^2 + 5x - 9$, calculate $f(1\frac{1}{10})$, by expanding $f(x+h)$ in powers of h, and then putting $x = 1$, $h = \frac{1}{10}$. [There will be but four terms in the expansion.]

31. In a similar manner, given $f(x) = x^3 - 6x^2 + 2x + 1$, calculate the value of $f(1.05)$.

32. Given $f(x) = x^5 - 4x^2 + 6$, calculate $f(1.1)$.

33. Show that $\log\left(x + \sqrt{x^2+4}\right) = \log 2 + \frac{x}{2} - \frac{1}{2}\frac{x^3}{3\cdot 2^3} + \frac{1\cdot 3}{2\cdot 4}\frac{x^5}{5\cdot 2^5} - \text{etc.}$ Find the interval of convergence. Use the series to find $\log\left(2 + 2\sqrt{2}\right)$.

34. Expansion in series can also be used to evaluate roots of trinomials. The binomial theorem can be used to get the series. Calculate $\sqrt[3]{1+x+x^2}$ for $x = .1$, which gives $\sqrt[3]{1.11}$.

35. In a similar manner calculate $1/\sqrt[6]{1.11}$ by use of the expansion of $1/\sqrt[5]{1+x+x^2}$.

36. Calculate $\sqrt[7]{5/6}$ by use of $(1+x)^{-\frac{1}{7}}$ and putting $x=\frac{1}{5}$.

37. Expand $1/(x-h)$ as a series in powers of $1/x$ and use it to calculate the reciprocal of 97.

EXERCISE LIX

Indeterminate Forms. Evaluation by Series. Forms which take the indeterminate form $0/0$ may be evaluated by expanding both numerator and denominator in powers of $x-a$, where $x=a$ is the value for which the form becomes indeterminate. If $x=0$ gives the indeterminate value, the series is simply a power series in x. If the series assumes the value ∞/∞, the same method is available, although it may be necessary to change the form before expansion; for example, when $x=\infty$ gives the indeterminate value, let $x=1/z$, expand in powers of z, and then let $z=0$.

Evaluate by expansion. (After expansion divide through by any common factor in x or $x-a$.) Two terms in each series are ample.

1. $\dfrac{\sin x}{x}$, $x=0$.

2. $\dfrac{\sin x}{e^x-1}$, $x=0$.

3. $\dfrac{1-\cos x}{1-e^x}$, $x=0$.

4. $\dfrac{x-\sin x}{x^3}$, $x=0$.

5. $\dfrac{\sin x}{\sin 3x}$, $x=0$.

6. $\dfrac{x\cos x-\sin x}{x^3}$, $x=0$.

7. $\dfrac{1-\cos x}{\sin^2 x}$, $x=0$.

8. $\dfrac{\sin 2x-2\sin x}{2e^x-2-2x}$, $x=0$.

9. $\dfrac{xe^x+e^x-1}{\sin x}$, $x=0$.

10. $\dfrac{\log(1+x)}{\sin 2x}$, $x=0$.

11. $\dfrac{\log x}{x^2-1}$, $x=1$.

12. $\dfrac{\sin \pi x}{\cos \pi x/2}$, $x=1$.

13. $\dfrac{\sin(x-1)}{\tan \pi x}$, $x=1$.

14. $\dfrac{x\cos x}{x-\sin x}$, $x=0$.

15. $\dfrac{\sin x-x}{x\sin x}$, $x=0$.

16. $\dfrac{x^2-\sin^2 x}{x^2\sin^2 x}$, $x=0$.

17. $\dfrac{\cos x-e^x}{\operatorname{arc}\sin x}$, $x=0$.

18. $\dfrac{1-\cos x}{\cos x\sin^2 x}$, $x=0$.

EXERCISE LX

Taylor's Theorem — Several Variables. The extension of Taylor's and Maclaurin's theorems to functions of several variables gives rise to the following formulas for two variables (which can be abridged by proper symbolism):

$$(1)\ f(x+h,\ y+k)=f(x,\ y)+h\frac{\partial f}{\partial x}+k\frac{\partial f}{\partial y}$$
$$+\frac{1}{2!}\left(h^2\frac{\partial^2 f}{\partial x^2}+2\,hk\frac{\partial^2 f}{\partial x\partial y}+k^2\frac{\partial^2 f}{\partial y^2}\right)$$
$$+\frac{1}{3!}\left(h^3\frac{\partial^3 f}{\partial x^3}+3\,h^2k\frac{\partial^3 f}{\partial x^2\partial y}+3\,hk^2\frac{\partial^3 f}{\partial x\partial y^2}+k^3\frac{\partial^3 f}{\partial y^3}\right)+\cdots.$$

$$(2)\ f(x,\ y)=f(x_0,\ y_0)+x\frac{\partial f}{\partial x_0}+y\frac{\partial f}{\partial y_0}$$
$$+\frac{1}{2!}\left(x^2\frac{\partial^2 f}{\partial x_0{}^2}+2\,xy\frac{\partial^2 f}{\partial x_0\partial y_0}+y^2\frac{\partial^2 f}{\partial y_0{}^2}\right)$$
$$+\frac{1}{3!}\left(x^3\frac{\partial^3 f}{\partial x_0{}^3}+3\,x^2y\frac{\partial^3 f}{\partial x_0{}^2\partial y_0}+3\,xy^2\frac{\partial^3 f}{\partial x_0\partial y_0{}^2}\right.$$
$$\left.+y^3\frac{\partial^3 f}{\partial y_0{}^3}\right)+\cdots,$$

where $f(x_0,\ y_0)$ means $f(0,\ 0)$ and $\partial f/\partial x_0$ means $\partial f/\partial x$ evaluated for *both x and y equal to zero*, and so for the other partial derivatives.

1. Given $f(x,\ y)=x^3+xy^2$, expand $f(x+1,\ y+2)$ by use of the extended Taylor's theorem.

2. Given $f(x,\ y)=x^3+2\,xy+y^2-x$, develop $f(x-1,\ y+1)$.

Expand the following in powers of x and y.

3. $e^{xy}\sin x$. 4. $a^x\log(1+y)$. 5. $\cos(xy)$. 6. $\sin x\sin y$.

7. $\cos x\cos y$. 8. Expand $\sin(x+y)$ and $\sin(x+h+y+k)$.

9. Expand $\cos[(x+h)(y+k)]$.

In the next five, expand $f(x+h,\ y+k)$ also, where $f(x,\ y)$ is the given function.

10. $e^x\sin y$. 12. $\sin x\cos y$. 14. $\text{arc}\tan(x/y)$.

11. $\log(e^x+e^y)$. 13. $x\log y$. 15. $\cos(x/y)$.

16. $e^y \cos\sqrt{x}$.
17. $\sin(x+y)\log(1+x)$.
18. $\sin y \operatorname{arc}\tan x$.
19. $e^{x+\sin y}$.
20. $2^x e^y$.
21. $\tan(xe^y)$.
22. $e^y/(1-x)$.
23. $\sin x/(y-e^y)$.
24. $\sin\left(y\sqrt{1-x^2}\right)$.
25. $\sqrt{1-x^2}\sin e^y$.
26. $e^{xy}\sin(x-y)$.
27. $e^x \sin y/(1-z)$.
28. $(x+y+z)^n$.
29. $e^y \log(1-x)\tan z$.
30. $\sin x \sin y \sin z$.

EXERCISE LXI

Maxima and Minima—Two Variables. If $F(x, y)$ has a relative maximum or minimum value, the following three conditions must be satisfied:

$$(1)\ \frac{\partial F}{\partial x}=0, \quad (2)\ \frac{\partial F}{\partial y}=0, \quad (3)\ \left(\frac{\partial^2 F}{\partial y \partial x}\right)^2 < \frac{\partial^2 F}{\partial x^2}\cdot\frac{\partial^2 F}{\partial y^2}.$$

At maximum points, $\partial^2F/\partial x^2 < 0$, or $\partial^2 F/\partial y^2 < 0$.

At minimum points, $\partial^2F/\partial x^2 > 0$, or $\partial^2 F/\partial y^2 > 0$.

Find the maximum and minimum values of the following functions:

1. $z = x^2 + y^2 + xy - 6x - 4y + 5$.
2. $z = x^2 - 6xy + y^3 + 3x + 6y$.
3. $z = 4xy + 1/x + 1/y$.
4. $z = x^2 + y^2 + xy - x - 5y$.
5. $z = \sin x + \sin y + \sin(x+y)$.
6. $z = xy^2(6 - x - y)^3$.
7. $z = \sin x \sin y \sin(\pi - x - y)$.
8. $z = x^3 - 6xy + y^3$.
9. $z = 7x^2 - 6xy + 3y^2 - 4x + 7y - 12$.
10. $z = x^3 + y^2 - 6xy - 39x + 18y + 20$.
11. $z = x^2 + 2xy + 2y^2 + 4x - 4y + 6$.
12. $z = 2xy - x^2 - 4y^2 + 2x + 6y + 12$.
13. $z = x^2 + xy + y^2 - 5x - 4y + 1$.

14. Show that $u = x^3y^2(1 - x - y)$ is a relative maximum when $x = \frac{1}{2}$, $y = \frac{1}{3}$.

15. Show that the general quadratic $ax^2 + by^2 + 2hxy + 2gx + 2fy + c$ has the maximum value $\begin{vmatrix} a & h & g \\ h & b & f \\ g & f & c \end{vmatrix} \div \begin{vmatrix} a & h \\ h & b \end{vmatrix}$.

16. The volume of a right parallelopiped is 64 cu. ft. What are its dimensions for minimum total surface?

17. Divide the number 18 into three parts so that the continued product is a maximum.

18. Prove that the sum of three positive quantities whose product is a constant is a maximum when the three quantities are all equal.

19. The sum of the sides of a triangle is 12. Show that the area is a maximum when the triangle is equilateral.

20. Divide a straight line of fixed length into three segments so that the sum of the squares of these may be a minimum.

21. Find the dimensions of the rectangular parallelopiped of maximum surface which can be inscribed in a given sphere.

22. By the former regulations of the parcel post the sum of length and girth of any parcel must not exceed 6 ft. Show that the most voluminous cylindrical package which could be sent was a cylinder 2 ft. long, with girth 4 ft., and hence of volume 2.546 cu. ft. Under the new regulations this sum has been increased to 7 ft. What is the increase allowed in the maximum cylindrical package?

23. If $a/x + b/y + c/z = 1$, where a, b, and c are all positive, what are the values of x, y, and z which make the sum $x + y + z$ a minimum.

24. If x, y, z are the lengths of the perpendiculars dropped from any point P to the three sides a, b, c, respectively, of a triangle of area Δ, show that the minimum value of $x^2 + y^2 + z^2$ is equal to $4\,\Delta^2/(a^2 + b^2 + c^2)$.

25. Show that $\sin x + \sin y + \cos(x + y)$ has a minimum when $x = y = 3\,\pi/2$ and a maximum when $x = y = \pi/6$.

26. An open box has the form of a rectangular parallelopiped. If the volume is fixed, what are the dimensions for minimum material?

27. Find the minimum value of $x^2y^3z^4$ if $2\,x + 3\,y + 4\,z = a$.

28. Show that the maximum inscribed triangle and the minimum circumscribed triangle for a fixed circle are both equilateral.

29. One angle of an inscribed quadrilateral is 120°. Show that for maximum area the remaining pair of opposite angles are both 90°.

30. Given two mirrors OXM_1 and OXM_2 meeting in a right angle along the edge OX. P_1 is a luminous point, and its image is reflected first from the point N_1 in the first mirror, then from a point N_2 in the second mirror, and is finally caught on a screen at P_2. Assuming $P_1N_1N_2P_2$ to lie in a plane perpendicular to OX, what are the positions of N_1 and N_2 if the path $P_1N_1N_2P_2$ is a minimum?

31. Given a triangle ABC. Find the position of a point D so that the sum of the squares of the perpendiculars drawn to the three sides may be a minimum.

32. Given a triangle ABC, find the position of a fourth point D so that the sum $AD + BD + CD$ may be a minimum. (Cf. Sohncke, "Sammlung," p. 148, for this and similar problems.)

CHAPTER X

FURTHER APPLICATIONS TO GEOMETRY

EXERCISE LXII

Envelopes. If an equation contains a parameter (that is, a general constant) in addition to the variables x and y, we get a system of lines or curves by assigning various numerical values to this parameter. The locus of the limiting positions of the points of intersection of "consecutive" members of this system, or family, is called the envelope of the system. It is tangent to every line or curve of the system. To get the equation of the envelope, we differentiate the given equation *with respect to the parameter*, then eliminate the parameter between the original equation and this first derivative. Find the envelopes of the following systems of straight lines. *Draw figures.*

1. $y = mx + m^2$.

2. $y = x/m + m^2$.

3. $y = m^2x - 2m^3$.

4. $y = m^2x + 1/m^2$.

5. $xt^2 + yt - 1 = 0$.

6. $xt^2 - 2yt + 4 = 0$.

7. $x/h + y/k = 1$, where $h/3 + k/2 = 1$.

Find the envelopes of the following systems of circles:

8. $(x - \alpha)^2 + y^2 - 4\alpha = 0$.

9. $x^2 + (y - t)^2 = t$.

10. $(x - t)^2 + (y - t)^2 = 4$.

11. $(x - t)^2 + (y - t)^2 = 2t$.

Find the envelopes of the following families of curves:

12. $tx^2 + t^2y = 1$.

13. $y = t^2(x - t)$.

14. $y = tx^2 + 1/t$.

15. $y^2 = tx^3 + t$.

16. Find the envelope of the family of circles whose diameters are the double ordinates of the curve $y^2 = x^3$.

17. A variable circle passing through the origin moves with its center on the curve $y = x^3$. Find the equation of the envelope of the system of circles.

18. The ellipses of a system have the same area, the same center and axes always lying in the coördinate axes. Show that the envelope consists of two rectangular hyperbolas.

19. A right triangle moves so that one leg always passes through the point (4, 0), while the vertex of the right angle, C, moves along the y-axis. Find the envelope of the other leg as C moves.

Hint. Take the distance OC as the parameter.

20. Given the ellipse $x^2 + 4y^2 = 16$. There is a family of ellipses, equal in size to the given ellipse, with their centers on the perimeter of this ellipse, and with the respective axes parallel to the coördinate axes. Find the envelope of the system.

21. Given the two concentric ellipses $x^2 + 4y^2 = 64$ and $x^2 + 4y^2 = 16$. Find the envelope of the polar lines of the points of the first ellipse with respect to the second.

NOTE. The polar line of (a, b) with respect to $x^2 + 4y^2 = 16$ is $ax + 4by = 16$.

22. Find the envelope of the chords joining the points of contact of all pairs of mutually perpendicular tangents to the ellipse $16x^2 + 9y^2 = 144$.

Hint. Show first that the locus of the intersections of these pairs of tangents is the director circle $x^2 + y^2 = 25$.

23. Find the envelope of all straight lines such that the sum of their intercepts is constant and equal to 6.

24. Find the envelope of the system of parabolas $(y - 4)^2 = k(x - k)$.

25. Find the envelope of the normals to the parabola $(y - 2)^2 = 8x$, and show that it is the evolute, the locus of the centers of curvature.

26. Find the envelope of the normals to the ellipse $9x^2 + 25y^2 = 225$.

27. Given the family of ellipses $b^2x^2 + a^2y^2 = a^2b^2$, with the restriction $a + b = 8$. Find the envelope of the family

28. Given the parabola $y^2 = 4x$. Find the envelope of the family of parabolas of the same size having their foci on the given parabola.

EXERCISE LXIII

Further Applications to Plane Curves. (A) ***Asymptotes to Algebraic Curves.*** There are two standard methods of calculating the equations of the asymptotes to algebraic curves, (1) the method of limiting intercepts, which is strictly speaking the method of the calculus, and (2) the algebraic method of analytic geometry. The former method is given in detail in almost any textbook on the calculus. By method (2), which is in general much simpler, we arrange the equation in powers of y. If the coefficient of the highest power of y is a function of x, the real root or roots of this function equated to zero give the vertical asymptotes. To find the oblique asymptotes, assume them to be of the form $y = mx + k$. Substitute $mx + k$ for y in the original equation. Equate to zero the coefficients of the two highest powers of x. Solve these two equations for the pair or pairs of values of m and k, say m_1 and k_1. Then $y = m_1x + k_1$ is an asymptote.

Find the equations of the real asymptotes to the following curves by either method:

1. $y^3 = x^2(x - a)$.

2. $2x^2 - 5xy + 3y^2 - 7x + 4y - 12 = 0$.

3. $x^3 - 3xy^2 - 2y^3 - 4x + 8y - 9 = 0$.

4. $y^2(x - 4) - x^3 - 4x^2 = 0$.

5. $x^3 - x^2y + y = 0$.

6. $xy^2 - x + 2y - 1 = 0$.

7. $xy^2 + yx^2 - 8 = 0$.

8. $x^2 - y^2 - 2x - 2y - 3 = 0$.

9. $x^3 - xy^2 - 4x^2 - 4y^2 = 0$.

10. $xy^2 - 2x^2y + x^3 - 4y = 0$.

11. $x^4 - y^4 - 4xy = 0$.

12. $x^3 = 8y^2 - y^3$.

13. $x^3 - xy^2 + 4y^2 = 0$.

14. $x^2y^2 = (x + 2)^2(1 + x^2)$.

15. $y^2(x - a) = x^3 + ax^2$.

16. $x^2(y - x) + xy + x^2 + x + 2 = 0$.

17. $(x + a)y^2 = (y + b)x^2$.

18. $x^3 + y^3 - x^2 - y^2 = 0$.

19. $x - 4xy^2 - 3x^2 + 12xy - 12y^2 + 8x + 2y + 4 = 0$.

20. $x^2y^2 - x^3y + x^2 - 4y^2 = 0$.

21. $(y - 2x)(y^2 - x^2) - a(y - x)^2 + 4a^2(x + y) = 0$.

22. $x^2y^2 + ax(x + y)^2 - 2a^2y^2 - a^4 = 0$.

23. $x^7 - x^3y^4 + a^4y^3 - ax^2y^4 = 0$.

24. $x^2(x - y)^2 - a^2(x^2 + y^2) = 0$.

25. $(x^2 - y^2)^2 - 4y^2 + y = 0$.

26. $x^5 + y^5 - 5ax^3y = 0$.

27. $x^5 - 5ax^2y^2 + y^5 = 0$.

28. $3x^3 - 8x^2y + 3xy^2 + 2y^3 + 3x^2 - 5xy - y^2 + 6x + 20 = 0$.

29. $y^3 - 3y^2x - yx^2 + 2x^3 + y^2 - 6xy + 5x^2 - 2y + 2x + 1 = 0$.

30. $x = (t - 8)/(t^2 - 4)$, $y = 3/(t^3 - 4t)$.

31. $x = t^2/(t - 2)$, $y = t^3/(t^2 - t - 2)$.

(B) ***Asymptotes to Polar Curves.*** For a polar curve to have an asymptote, there must be a value of θ which makes the radius vector ∞ and the polar subtangent finite. Hence to find the polar asymptotes we consider ρ and $\rho^2 d\theta/d\rho$, and if any finite value θ_1 of θ makes ρ infinite and $\rho^2 d\theta/d\rho$ finite, there is an asymptote parallel to the line $\theta = \theta_1$ and at a distance d from it, d being the finite value of the subtangent. Determine the asymptotes to the following:

1. $\rho = 4/(1 - 2\cos\theta)$.

2. $\rho = 2/(1 + 2\sin\theta)$.

3. $\rho\cos 2\theta = a$.

4. $\rho\sin 4\theta = a\sin 3\theta$.

5. $\rho^2\theta = \pi$.

6. $\rho = 4(\sec\theta + \tan\theta)$.

7. $\rho = a\sec\theta + b\tan\theta$.

8. $\rho^2\sin 2\theta = 4$.

9. $\rho\cos 2\theta = 4(1 + \sin 2\theta)$.

10. $\rho^2\sin\theta = a^2\cos 2\theta$.

11. $\rho^2\cos\theta = a^2\sin 3\theta$.

12. $\rho\sin\theta = a(\sin 2\theta + k)$.

13. $\rho\cos\theta = 4\sin^2\theta$.

14. $\rho\cos\theta = 4\cos 2\theta$.

15. $\rho = 2\sin\theta/(\tan\theta - 1)$.

16. $\rho = 2\theta/\sin\theta$.

17. $\rho\cos\theta = 4\cos 2\theta + 2\cos\theta$

18. $\rho\sin\theta = 2\cos 2\theta$.

19. $\rho\sin\theta = \cos 3\theta$.

20. $\rho = \pi/\theta$.

21. $\rho = 4(1 + \sin 2\theta)/\cos 2\theta$

22. $\rho = 2a\theta/(2\theta - 1)$.

23. $\rho = 1 - \sqrt{2 - \csc\theta}$.

24. $\rho\cos m\theta = a\cos n\theta$.

25. $\rho = a(\theta + \cos\theta)/(\theta + \sin\theta)$.

26. $\rho = a\theta^2/(\theta^2 - 1)$.

27. $(\rho - a)\sin\theta = b$.

28. $\rho = a\sec m\theta + b\tan m\theta$.

29. $\rho^2\sin(\theta - \alpha) + a\rho\sin(\theta - 2\alpha) + a^2 = 0$.

30. $\rho = 4(1 + 1/\cos 2\theta)$.

(C) ***Singular Points on Plane Curves.*** Maximum and minimum points and points of inflection have been treated in earlier exercises. The conditions for the principal point singularities are as follows:

For any singularity given below: $\partial f/\partial x = 0$, $\partial f/\partial y = 0$, at the point. Further conditions are

(a) For a real double point: $\left(\frac{\partial^2 f}{\partial x \partial y}\right)^2 - \frac{\partial^2 f}{\partial x^2}\cdot\frac{\partial^2 f}{\partial y^2} > 0$.

(b) For a cusp: $\left(\frac{\partial^2 f}{\partial x \partial y}\right)^2 - \frac{\partial^2 f}{\partial x^2} \cdot \frac{\partial^2 f}{\partial y^2} = 0.$

(c) For an isolated or conjugate point: $\left(\frac{\partial^2 f}{\partial x \partial y}\right)^2 - \frac{\partial^2 f}{\partial x^2} \cdot \frac{\partial^2 f}{\partial y^2} < 0.$

For a discussion of more complex singularities the student should consult a book on plane curves.

Investigate the following curves for singular points:

1. $x^3 + y^3 - x^2 - y^2 = 0.$

2. $x^3 - y^3 + (2y - x)^2 = 0.$

3. $x^4 - 2x^2y - 2xy^2 + 4y^2 = 0.$

4. $x^3 + y^3 - 3axy = 0$

5. $x(x-a)^2 + y^2(x - 2a) = 0.$

6. $ay^2 - x^3 + bx^2 = 0.$

7. $x^3 + 9x^2 - y^2 + 24x + 6y = 0.$

8. $x^3 - 4x^2 - 3y^2 + 12y - 12 = 0.$

9. $x^3 + xy^2 - 2x^2 - 2y + 2 = 0.$

10. $x^4 + 12x^3 - 6y^3 + 36x^2 + 27y^2 - 81 = 0.$

11. $x^4 + y^4 - 2x^2 - 2y^2 + 1 = 0.$ Note the number of singular points.

12. $x^4 + 4x^2 - 4y^2 = 0.$

13. $y^2 - x^3 - x^2 = 0.$

14. $x^4 + y^4 - 2y^2 - 8x^2 + 16 = 0.$

15. $x^4 - 4y^3 - 12y^2 - 8x^2 + 16 = 0.$

16. $(y-2)^2 = (x-3)^5.$

17. $5y^2 + 10y - x^3 + 2x^2 + 5 = 0.$

18. $x^2 - 2xy + y^2 - x^3 = 0.$

19. $4y^2 - x^3 + 2x^2 = 0.$

20. $y^2 = (2-x)^2(1-x).$

21. $3x^2 + 4y^2 - (x^2 + y^2)^2.$

22. $4x^2 - y^2 - x^3 = 0.$

23. $y = 2x^2 + x^{5/2}.$

24. $x^4 - ax^2y + axy^2 + \frac{1}{4}a^2y^2 = 0.$

25. $x^5 - 4ay^4 + 2ax^3y + a^2xy^2 = 0.$

26. $(x-y)^3 + a(x^2 - y^2) + a^2y = 0.$

27. $(x^2 + y^2)^2 + a^2(y^2 - x^2) = 0.$

28. $x^{2/3} + y^{2/3} = a^{2/3}.$

29. $x^4 - 4ax^3 + 4a^2x^3 - b^2y^2 + 2b^3y - a^4 - b^4 = 0.$

30. $x^5 + bx^4 - a^3y^2 = 0.$

31. $x^4 + y^4 - ay^3 - 2ax^2y + axy^2 + a^2y^2 = 0.$

32. $y^2(a^2 + x^2) = x^2(a^2 - x^2).$ Show that the origin is a node and that the nodal tangents bisect the angles made by the axes.

33. $x = t^2,\ y = t - \frac{1}{3}t^3.$ Find (a) radius of curvature at origin, (b) coördinates of maximum and minimum points, (c) coördinates of double point.

EXERCISE LXIV

Space Geometry. Curves. When a curve is given parametrically, the direction cosines of the tangent line to the curve at any point are proportional to the values of dx/dt, dy/dt, dz/dt, at the point. Call these values a, b, c. Then the tangent line is

$$(x-x_1)/a=(y-y_1)/b=(z-z_1)/c,$$

and the normal plane is

$$a(x-x_1)+b(y-y_1)+c(z-z_1)=0.$$

In case, however, the curve is given as the intersection of two surfaces, the direction numbers a, b, c of the tangent line are readily found. Since the parameter t may be x, obviously $a:b:c=dx/dx:dy/dx:dz/dx=1:dy/dx:dz/dx$. So it merely remains to find dy/dx and dz/dx at the required point. To do this differentiate both equations separately with respect to x, and then in these two equations replace x, y, z by the coördinates of the point in question. This gives two simple equations to solve for dy/dx and dz/dx. Call these values A and B. Then the equation of the tangent line is

$$(x-x_1)/1=(y-y_1)/A=(z-z_1)/B$$

and of the normal plane is

$$(x-x_1)+A(y-y_1)+B(z-z_1)=0.$$

In the second case, the two-plane form of the equations of the tangent line is found by getting the equation of the tangent plane to each surface at the given point. The intersection of these two planes is obviously the required tangent line. The first method seems to the author desirable, as it gives at once the direction numbers of the desired tangent line. See next exercise for the method of finding the equation of the tangent plane to a surface.

Find the equations of the tangent line and the equation of the normal plane to each of the following curves at the point indicated:

1. $x=t-3$, $y=t^2+1$, $z=t^2$, at $t=2$.
2. $x=t^2-1$, $y=t+t^3$, $z=1-2t$, at $t=1$.

3. $x = t^2$, $y = t^3$, $z = t^2 + 4$, at $t = 1$.
4. $x = 2t$, $y = 2/t$, $z = t^2$, at $t = 2$.
5. $x = 3 - t^2$, $y = t^3 - 1$, $z = 4/t$, at $t = 1$.
6. $x = 2t^2 - 3$, $y = 4/t$, $z = 3t$, at $t = 2$.
7. $x = t^2$, $y = 1/t$, $z = \sin \pi t/2$, at $t = 1$.
8. $x = \sin t$, $y = \cos t$, $z = \sec t$, at $t = 0$.
9. $x = 2 \sin t$, $y = 3 \cos t$, $z = 2t/\pi$, at $t = \pi/2$.
10. $x = t/(1 + t)$, $y = (1 + t)/t$, $z = t^2$, at $t = 1$.
11. $x^2 + y^2 + z^2 = 49$, $x^2 + y^2 = 13$, at $(3, 2, -6)$.
12. $z = x^2 + y^2 - 1$, $3x^2 + 2y^2 + z^2 = 30$, at $(2, 1, 4)$.
13. $x^2 - y^2 - z^2 = 1$, $x^2 - y^2 + z^2 = 9$, at $(3, 2, 2)$.
14. $x^2 + y^2 - z^2 = 16$, $x^2 + 4y^2 + 4z^2 = 84$, at $(2, 4, 2)$.
15. $x^2 + y^2 + 3z^2 = 32$, $2x^2 + y^2 - z^2 = 0$, at $(2, 1, 3)$.
16. $x^2 + 4y^2 - 4z^2 = 0$, $2x + y + z - 24 = 0$, at $(8, 3, 5)$.
17. $2x + 3y + 2z = 9$, $x^2 + 2y^2 + 2z^2 = 8$, at $(2, 1, 1)$.

EXERCISE LXV

Space Geometry. Surfaces. Since the direction cosines of the normal to a surface $F(x, y, z) = 0$ at any point are proportional to $\partial F/\partial x$, $\partial F/\partial y$, and $\partial F/\partial z$ at that point, it is very easy to write down the equation of the tangent plane to the surface as well as the equations of the normal line. For if the values of the above three partial derivatives are A, B, and C, the equations are

tangent plane, $A(x - x_1) + B(y - y_1) + C(z - z_1) = 0$;
normal line, $(x - x_1)/A = (y - y_1)/B = (z - z_1)/C$.

If a surface and a curve intersect, the angle between them is defined as the angle between the normal line to the surface and the tangent line to the curve at their common point, and can now be readily found. Find the equation of the tangent plane and the equations of the normal line to the following surfaces:

1. $x^2 + y^2 + z^2 = 49$, at $(6, 2, 3)$.
2. $z = x^2 + y^2 - 1$, at $(2, 1, 4)$.
3. $x^2 - y^2 - z^2 = 1$, at $(3, 2, 2)$.
4. $x^2 + y^2 - z^2 = 25$, at $(5, 5, 5)$.
5. $2x^2 + 3y^2 + 4z^2 = 6$, at $(1, 1, 1/2)$.
6. $x + y - z^2 = 3$, at $(3, 4, 2)$.

7. $x^2 + xy^2 + y^3 + z + 1 = 0$, at $(2, -3, 4)$.

8. $x^2 + 2xy + y^2 + z - 7 = 0$, at $(1, -2, 6)$.

9. $x^2y^2 + xz - 2y^3 - 10 = 0$, at $(2, 1, 4)$.

10. $x^2 + 4y^2 - 4z = 0$, at $(2, 2, 5)$.

11. The surface $x^2 - 4y^2 - 4z = 0$ is cut by the curve $x = t^2/2$, $y = 4/t$, $z = (t - 2t^2)/2$, at the point $(2, 2, -3)$. What is the angle of intersection?

12. The surface $x^2 + y^2 + 3z^2 = 25$ and the curve $x = 2t$, $y = 3/t$, $z = -2t^2$ intersect at the point given by $t = 1$ on the curve. What is the angle of intersection?

13. The ellipsoid $x^2 + 2y^2 + 3z^2 = 20$ and the skew curve $x = 3(t^2 + 1)/2$, $y = t^4 + 1$, $z = t^3$ meet in the point $(3, 2, 1)$. Find the equation of the tangent plane to the surface and also of the normal plane to the curve at this common point. What is the angle between the planes? What is your conclusion?

14. The surfaces $x^2y^2 + 2x + z^3 = 16$ and $3x^2 + y^2 - 2z = 9$ have the point $(2, 1, 2)$ lying on their curve of intersection. What are the equations of the respective tangent planes to the two surfaces at this point? What is the angle between the two normal lines?

15. Given the surfaces $x^2 + 3y^2 + 2z^2 = 9$ and $x^2 + y^2 + z^2 - 8x - 8y - 6z + 24 = 0$, show that they are tangent to each other at the point $(2, 1, 1)$.

16. Given the paraboloid $3x^2 + 2y^2 - 2z = 1$ and the sphere $x^2 + y^2 + z^2 - 4y - 2z + 2 = 0$, show that they cut orthogonally at the point $(1, 1, 2)$ by finding the angle between the respective tangent planes at this point.

CHAPTER XI

INTEGRAL CALCULUS. SIMPLE FORMAL INTEGRATION

EXERCISE LXVI

The Power Form. This most important formula is

$$\int v^n dv = \frac{v^{n+1}}{n+1} + C.$$

It holds for n any real value except $n = -1$. If the denominator consists of a quantity raised to a power, it should be brought up with a negative exponent. The student should observe carefully what part of the integrand is represented by v in the formula, and then see that dv,* the differential of v, is present, and not merely dx. In fact, it is not amiss to carry out the actual substitution, even in the simpler cases, until perfectly familiar with the process. It is best to begin by putting any constant factor of the integrand outside the integral sign. Then observe whether the differential of v is present. If a constant factor is lacking, it may be supplied, but we must at the same time compensate for this by dividing outside the integral sign by the same constant. Note carefully that we can *not* supply any lacking term if it contains a variable and compensate for it outside the integral sign. These principles are general and do not apply to this exercise alone. They are shown step by step in these illustrative examples. The actual integration comes last always. All answers should be verified by differentiating them, thus obtaining the integrand when correct.

* The differential of one variable (usually dx in these examples) is always present. It is, in fact, a part of the integral sign designating the variable with respect to which we are integrating

I. $\int 3\sqrt{(x^2+6)}\,x\,dx = 3\int (x^2+6)^{\frac{1}{2}}x\,dx = \frac{3}{2}\int (x^2+6)^{\frac{1}{2}}\cdot 2\,x\,dx$

$$= \tfrac{3}{2}(x^2+6)^{\frac{3}{2}}/\tfrac{3}{2} = (x^2+6)^{\frac{3}{2}} + C.$$

II. $\int 3(x^2+1)^2 dx = 3\int (x^2+1)^2 dx$ (here we cannot "compensate," so must try some other device, namely, expansion by the binomial theorem) $= 3\int (x^4 + 2\,x^2 + 1)\,dx$

$$= 3\left(\int x^4 dx + 2\int x^2 dx + \int dx\right)$$

$$= 3(x^5/5 + 2\,x^3/3 + x) + C.$$

Be careful of signs in compensating.

1. $\int x^7 dx.$
2. $\int x^{\frac{5}{3}} dx.$
3. $\int x^{\frac{1}{3}} dx.$
4. $\int (1/x^5)\,dx.$
5. $\int \sqrt{x}\,dx.$
6. $\int (1/\sqrt{x})\,dx.$
7. $\int (\sqrt{x})^5 dx.$
8. $\int (1/\sqrt{x})^3 dx.$
9. $\int \sqrt{2\,x}\,dx.$
10. $\int (4/\sqrt{2\,x})\,dx.$
11. $\int (\sqrt{3}/x^2)\,dx.$
12. $\int (2\,x^3/\sqrt{x})\,dx.$
13. $\int (1+\sqrt{x})\,dx.$
14. $\int (\sqrt{x}+x^2)\,dx.$
15. $\int (x - 1/\sqrt{x})\,dx.$
16. $\int (x^{\frac{7}{3}} - x^{-\frac{1}{3}})\,dx.$
17. $\int (1-x)^3 dx.$
18. $\int (1+\sqrt{x})^2 dx.$
19. $\int (1-1/x^{\frac{1}{3}})^2 dx.$
20. $\int (1-1/x^2)^2 dx.$
21. $\int (y^2+1/y^2)^2 dy.$
22. $\int (z^{\frac{1}{3}} - z^{-\frac{1}{3}})^2 dz.$
23. $\int (\sqrt{s} + s^{\frac{1}{3}})\,ds.$
24. $\int (t^{\frac{1}{5}} - 1/\sqrt{t})\,dt.$
25. $\int \sqrt{x}(1-x^2)\,dx.$
26. $\int \sqrt{x}(1-x)^2 dx.$
27. $\int x^2(1-\sqrt{x})\,dx.$
28. $\int x(1-\sqrt{x})^2 dx.$
29. $\int (1-\sqrt{z})^2/z^{\frac{1}{3}}\cdot dz.$
30. $\int (z-1)^3/\sqrt{z}\cdot dz.$
31. $\int (\sqrt{x}+x^{\frac{1}{3}})^2/x\cdot dx.$
32. $\int (\sqrt{y}+\sqrt[4]{y})^2/y\cdot dy.$
33. $\int (3-m)^4 dm.$
34. $\int (3\,n+2)^3 dn.$
35. $\int (3-5\,x)^{-7} dx.$
36. $\int 7/(1+2\,x)^3\cdot dx.$
37. $\int \sqrt{4-t}\,dt.$
38. $\int \sqrt{2\,s+1}\,ds.$
39. $\int (1/\sqrt{1+3\theta})\,d\theta.$
40. $\int (1/\sqrt{3-2\,B})\,dB.$
41. $\int s(s^2+1)^7 ds.$
42. $\int t^2(2-t^3)^3 dt.$

43. $\int x^2/(3+2x^3)^2 \cdot dx.$ 45. $\int 5x\sqrt{1-2x^2}\,dx.$ 47. $\int (2x^{\frac{3}{2}}+1)^{\frac{2}{3}}\sqrt{x}\,dx.$

44. $\int x^3/(5-x^4)^3 \cdot dx.$ 46. $\int 5x^3/\sqrt{1+x^4}\cdot dx.$ 48. $\int (2-3x^{\frac{4}{3}})^{\frac{1}{3}}x^{\frac{1}{3}}\,dx.$

49. $\int x^{\frac{2}{3}}/(2x^{\frac{5}{3}}+5)^{\frac{1}{3}} \cdot dx.$

50. $\int x^{\frac{3}{4}}/(1-x^{\frac{7}{4}})^{\frac{2}{3}} \cdot dx.$

51. $\int (x+1)/\sqrt{x^2+2x} \cdot dx.$

52. $\int (3x^2-6x)^3(x-1)\,dx.$

53. $\int (x^3+3x-1)(x^2+1)\,dx.$

54. $\int (x+1)/\sqrt{x^2+2x-1} \cdot dx.$

55. $\int \sqrt{(1-\sqrt{x})/x} \cdot dx.$

56. $\int \sqrt{1/x(1-\sqrt{x})} \cdot dx.$

57. $\int (1+e^{3x})^2 e^{3x}\,dx.$

58. $\int \sqrt{1-e^x}\,e^x\,dx.$

59. $\int (2+\log x)/x \cdot dx.$

60. $\int \sqrt{\log x - 3}/x \cdot dx.$

61. $\int \sin^2\frac{x}{2}\cos\frac{x}{2}\,dx.$

62. $\int \tan 2x \sec^2 2x\,dx.$

63. $\int \sin 3x/\sqrt{\cos 3x + 5}\,dx.$

64. $\int [\sec x/(1+\tan x)]^2\,dx.$

EXERCISE LXVII

The Logarithm Form. The general form is

$$\int \frac{dv}{v} = \log v + C.$$

The student should observe that this is the special case ($n=-1$) of the integral of the previous exercise where that formula fails. Hence if the integrand is a fraction, the student should *first notice whether the numerator is merely a multiple of the derivative of the denominator* (or can be so written). If it is, the present formula applies. If it is not, the denominator is brought up with a negative exponent and the power form is tried.

1. $\int \frac{dx}{x+1}.$ 4. $\int \frac{x\,dx}{2-7x^2}.$ 7. $\int \frac{5x^{\frac{3}{2}}\,dx}{x^{\frac{5}{2}}+2}.$ 10. $\int \frac{\sqrt{x}\,dx}{1+x^{\frac{3}{2}}}.$

2. $\int \frac{dx}{1+3x}.$ 5. $\int \frac{2x^2\,dx}{x^3+1}.$ 8. $\int \frac{8x^{\frac{5}{3}}\,dx}{3-2x^{\frac{8}{3}}}.$ 11. $\int \frac{dx}{(2+\sqrt{x})\sqrt{x}}.$

3. $\int \frac{dx}{1-2x}.$ 6. $\int \frac{7x^3\,dx}{2-5x^4}.$ 9. $\int \frac{4\,dx}{x^{\frac{3}{4}}(5+x^{\frac{1}{4}})}.$ 12. $\int \frac{(3\sqrt{x}+2)\,dx}{(x+\sqrt{x})\sqrt{x}}.$

13. $\int \frac{(x-1)\,dx}{x^2-2x+3}$.

14. $\int \frac{(1-x^2)\,dx}{6x-2x^3}$.

15. $\int \frac{(x-x^3)\,dx}{x^4-2x^2+5}$.

16. $\int \frac{e^{2x}dx}{2+e^{2x}}$.

17. $\int \frac{dx}{e^x(3+e^{-x})}$.

18. $\int \frac{3^{3x}dx}{1-3^{3x}}$.

19. $\int \frac{dx}{x(1+\log x)}$.

20. $\int \frac{dx}{1+\log e^x}$.

21. $\int \frac{\log x\,dx}{x(1-\log^2 x)}$.

22. $\int \frac{\cos 2x\,dx}{\sin 2x}$.

23. $\int \frac{\sin 2x\,dx}{1+\sin^2 x}$.

24. $\int \frac{3\cos 5x\,dx}{1-2\sin 5x}$.

25. $\int \frac{\cot x\,dx}{\log \sin x}$.

26. $\int \frac{(1+\cos x)\,dx}{x+\sin x}$.

27. $\int \frac{(x-\sin 2x)\,dx}{x^2+\cos 2x}$.

28. $\int \frac{(1+\tan x)\,dx}{x+\log \sec x}$.

29. $\int \frac{dx}{\sin 2x \log \tan x}$.

30. $\int \frac{dx}{(1+x^2)\,\text{arc}\tan x}$.

EXERCISE LXVIII

Exponential Functions. The formulas are simply the inverse of the corresponding differentiation formulas.

$$\int a^v dv = a^v/\log a + C, \qquad \int e^v dv = e^v + C.$$

The student is again reminded of the importance of dv, its presence being necessary before the formulas can be applied. The above integral should not be confused with the power form $\int v^a dv$.

1. $\int 2^x dx$.
2. $\int 5^{2x} dx$.
3. $\int (1/3^x)\,dx$.
4. $\int (1/e^x)\,dx$.
5. $\int (1/2^{5x})\,dx$.
6. $\int \pi^{3x} dx$.
7. $\int e^{(5x+1)} dx$.
8. $\int e^{(1-3x)} dx$.
9. $\int [1/e^{(1-2x)}]\,dx$.
10. $\int xe^{x^2} dx$.
11. $\int (e^{\sqrt{x}}/\sqrt{x})\,dx$.
12. $\int (e^{\frac{1}{x}}/x^2)\,dx$.
13. $\int e^{\sin 2x}\cos 2x\,dx$.
14. $\int \sin x/e^{\cos x}\cdot dx$.
15. $\int (\sec x\cdot e^{\tan x})^2 dx$.
16. $\int (e^{\cot x}/\sin^2 x)\,dx$.
17. $\int (e^{\log 5x}/x)\,dx$.
18. $\int [e^{\log(1-4x)}/(1-4x)]\,dx$.
19. $\int (\tan x/e^{\log \cos x})\,dx$.
20. $\int \cot x\cdot e^{\log \sin^2 x} dx$.
21. $\int (e^x + e^{3x})^2 dx$.

22. $\int (e^{2x} + e^{-2x})^2 \, dx.$ 25. $\int (e^{x/2} + e)^3 \, dx.$ 28. $\int (e^{x/5} + e^{5/2})^2 \, dx.$

23. $\int (e^{2x} + e^{x/2})^2 \, dx.$ 26. $\int (e^x - \sqrt{e})^2 \, dx.$ 29. $\int \left[(e^{2/x} + \sqrt{x})/x^2\right] dx.$

24. $\int (e^x + e^{x/3})^3 \, dx.$ 27. $\int (e^{3x} + 1/e)^2 \, dx.$ 30. $\int \left[(e^{\sqrt{x}} + x)/\sqrt{x}\right] dx.$

*31. $\int [e^{2x}/(e^x + 1)] \, dx.$ 33. $\int [e^x/(e^{x/2} + 2)] \, dx.$

32. $\int [e^{3x}/(1 - e^x)] \, dx.$ 34. $\int [(e^x + 1)^2/(e^x + 2)] \, dx.$

EXERCISE LXIX

Integration of the Fundamental Trigonometric Functions. In addition to memorizing the integrals of the six fundamental functions,

$$\int \sin v \, dv = -\cos v + C, \qquad \int \cos v \, dv = \sin v + C,$$

$$\int \tan v \, dv = -\log \cos v + C, \text{ or } \log \sec v + C,$$

$$\int \cot v \, dv = \log \sin v + C, \qquad \int \sec v \, dv = \log(\sec v + \tan v) + C,$$

$$\int \csc v \, dv = \log(\csc v - \cot v) + C,$$

and those which follow from the formulas for differentiation,

$$\int \sec^2 v \, dv = \tan v + C, \qquad \int \sec v \tan v \, dv = \sec v + C,$$

$$\int \csc^2 v \, dv = -\cot v + C, \qquad \int \csc v \cot v \, dv = -\csc v + C;$$

the student should have at his command the various simple relations existing among the six trigonometric functions. By means of these, the following expressions, if not already standard integrals, are easily reduced to such.

1. $\int \sin 2x \, dx.$ 3. $\int \sec 5x \, dx.$ 5. $\int \csc (x/2) \, dx.$

2. $\int \tan 3x \, dx.$ 4. $\int \cos (x/3) \, dx.$ 6. $\int \cot (x/7) \, dx.$

* **Reduce examples 31–34 to mixed numbers before integrating.**

7. $\int \tan(x/\sqrt{2})\,dx.$ 12. $\int \sec(3x+2)\,dx.$ 17. $\int x\sin(1-x^2)\,dx.$

8. $\int \sin \pi x\,dx.$ 13. $\int \sec^2 5x\,dx.$ 18. $\int (\cos\sqrt{x}/\sqrt{x})\,dx.$

9. $\int \csc 2\pi x\,dx.$ 14. $\int \sec 2x \tan 2x\,dx.$ 19. $\int (x\sec x^3)^2\,dx.$

10. $\int \cos(1+2x)\,dx.$ 15. $\int \csc^2(x/2)\,dx.$ 20. $\int [\csc(1/x)/x]^2\,dx.$

11. $\int \cot(1-x)\,dx.$ 16. $\int x\cos x^2\,dx.$ 21. $\int (\sec\sqrt{x}/\sqrt[4]{x})^2\,dx.$

22. $\int \frac{dx}{\sec 3x}.$ 23. $\int \frac{dx}{\cos 2x}.$ 24. $\int \frac{dx}{\cos^2 3x}.$ 25. $\int \tan^2 x/2\,dx.$

26. $\int \cot^2 3x\,dx.$ 31. $\int \frac{\sin^2 2x\,dx}{\cos 2x}.$ 36. $\int \frac{2\tan x\,dx}{\sin 2x}.$

27. $\int \left(\frac{\tan 1/x}{x}\right)^2 dx.$ 32. $\int \frac{\cos^2 3x\,dx}{\sin 3x}.$ 37. $\int \frac{e^x\,dx}{\cot e^x}.$

28. $\int \frac{\tan 3x\,dx}{\sec 3x}.$ 33. $\int \frac{\tan^2 x/2\,dx}{\sec x/2}.$ 38. $\int \frac{\sin e^{-x}\,dx}{e^x}.$

29. $\int \frac{\tan^2 2x\,dx}{\sin^2 2x}.$ 34. $\int \frac{\sin 2x\,dx}{2\sin x}.$ 39. $\int \frac{\tan \pi x\,dx}{\cot \pi x}.$

30. $\int \frac{\sin 5x\,dx}{\cos^2 5x}.$ 35. $\int \frac{\cos 2x\,dx}{\cos^2 x}.$ 40. $\int \tan 2x\cos 2x\,dx.$

41. $\int \tan 3x \sec 3x\,dx.$ 46. $\int (\tan x/2 + \cot x/2)^2\,dx.$

42. $\int (1+\sec 2x)^2\,dx.$ 47. $\int (\sec x/3 + \tan x/3)^2\,dx.$

43. $\int (1-\tan 3x)^2\,dx.$ 48. $\int (\sec 3x + \cot 3x)^2\,dx.$

44. $\int (\sin 2x + \cos 2x)^2\,dx.$ 49. $\int (\sec 2x + \csc 2x)^2\,dx.$

45. $\int (\sin 3x - \cos 3x)^2\,dx.$ *50. $\int \frac{dx}{1+\cos 2x}.$

51. $\int \frac{dx}{1-\sin 3x}.$ 52. $\int \frac{dx}{\sec 3x + 1}.$ 53. $\int \frac{dx}{\csc 2x - 1}.$

* Recall that $1+\cos 2x = 2\cos^2 x$. Similarly for the remaining examples the form must be radically changed before integration.

EXERCISE LXX

Constant over Quadratic. The examples here are of the form $\int (k/Q)\,dx$. The letter Q denotes the general quadratic function $ax^2 + bx + c$, and this integral should not be confused with either $\int (k/x)\,dx$ or $\int (k/Q)\,dQ$. In case Q is a perfect square the integral reduces to the power form of Exercise LXVI. Otherwise the quadratic can always be written as the sum or difference of two squares, and hence the integral takes one of the following forms, as the constant k can be taken outside the integral sign:

$$\int \frac{dv}{v^2 + a^2} = \frac{1}{a} \arctan \frac{v}{a} + C, \qquad \int \frac{dv}{v^2 - a^2} = \frac{1}{2a} \log \frac{v-a}{v+a} + C.$$

The student should understand clearly just what part of the expression represents v, and assure himself that dv, and not merely dx, is present before he applies either of the two formulas. (If Q has rational linear factors, it can be integrated also by partial fractions, as in Exercise LXXXI.)

These formulas may also be applied to the integration of other expressions wherein v is not necessarily linear or algebraic, as illustrated by examples 37–48 of this list. In the more involved cases it is sometimes desirable to carry out the transformation and reduce the integral to one of the two forms above. For instance, in example 48 let $v = \tan x$, $dv = \sec^2 x\,dx$, and we have

$$\int [1/(v^2+1)]\,dv = \arctan v + C = \arctan(\tan x) + C.$$

1. $\int \frac{dx}{x^2+1}$.
2. $\int \frac{dx}{x^2-1}$.
3. $\int \frac{dx}{1-x^2}$.
4. $\int \frac{dx}{x^2+4}$.
5. $\int \frac{dx}{9x^2-1}$.
6. $\int \frac{dx}{4-9x^2}$.
7. $\int \frac{dx}{x^2-5}$.
8. $\int \frac{dx}{2x^2+1}$.
9. $\int \frac{dx}{x^2+\sqrt{3}}$.
10. $\int \frac{3\,dz}{6+5z^2}$.
11. $\int \frac{d\theta}{7\theta^2+9}$.
12. $\int \frac{2\,dm}{3-8m^2}$.

13. $\int\frac{dx}{x^2+2x}$.

14. $\int\frac{dx}{4x-x^2}$.

15. $\int\frac{dx}{x^2-3x}$.

16. $\int\frac{dx}{5x-x^2}$.

17. $\int\frac{dx}{x^2+4x+3}$.

18. $\int\frac{dx}{2x-10-x^2}$.

19. $\int\frac{3\,dx}{x^2-8x+25}$.

20. $\int\frac{5\,dx}{x^2+12x+11}$.

21. $\int\frac{dx}{4x-x^2-2}$.

22. $\int\frac{dx}{x^2-6x+12}$.

23. $\int\frac{dx}{x^2-x-1}$.

24. $\int\frac{dx}{x^2+3x+3}$.

25. $\int\frac{dx}{4x^2+4x+5}$.

26. $\int\frac{dx}{9x^2-6x-8}$.

27. $\int\frac{5\,dy}{9y^2+12y-21}$.

28. $\int\frac{4\,dz}{4z^2+12z+5}$.

29. $\int\frac{12\,dt}{9t^2-6t-5}$.

30. $\int\frac{8\,dt}{16t^2+8t+9}$.

31. $\int\frac{dx}{9x^2+3x+1}$.

32. $\int\frac{dx}{9x^2-3x-1}$.

33. $\int\frac{2\,dx}{3x^2+4x+5}$.

34. $\int\frac{4\,dx}{5x^2+6x+7}$.

35. $\int\frac{4\,dx}{5x^2-7x-6}$.

36. $\int\frac{2\,dx}{3x^2-5x+4}$.

37. $\int\frac{x\,dx}{x^4+2x^2+2}$.

38. $\int\frac{x^2\,dx}{x^6-6x^3+5}$.

39. $\int\frac{dx}{\sqrt{x}(x-1)}$.

40. $\int\frac{dx}{x^{\frac{2}{3}}(x^{\frac{2}{3}}+1)}$.

41. $\int\frac{e^x\,dx}{e^{2x}+4}$.

42. $\int\frac{e^{x/2}\,dx}{3-e^x}$.

43. $\int\frac{dx}{x(\log^2 x+1)}$.

44. $\int\frac{dx}{x(\log^2 x-1)}$.

45. $\int\frac{\cos 3x\,dx}{\sin^2 3x+9}$.

46. $\int\frac{\sin 2x\,dx}{4-\cos^2 2x}$.

47. $\int\frac{\sec^2 x\,dx}{\tan^2 x-1}$.

48. $\int\frac{\sec^2 x\,dx}{\tan^2 x+1}$.

EXERCISE LXXI

Constant over the Square Root of a Quadratic, or $\int (k/\sqrt{Q})\,dx$.

This higher irrationality is classed under one of two standard forms, depending on the form assumed by Q when written as the algebraic sum of two squares.

$$\int\frac{dv}{\sqrt{a^2-v^2}}=\text{arc sin}\,\frac{v}{a}+C,$$

$$\int\frac{dv}{\sqrt{v^2\pm a^2}}=\log\left(v+\sqrt{v^2\pm a^2}\right)+C.$$

The student should *guard against the common error of writing* $\int\frac{dv}{\sqrt{a^2-v^2}}=-\int\frac{dv}{\sqrt{v^2-a^2}}$, and should observe therefore that *when the coefficient of x^2 in Q is negative, the above integral is invariably an arc sine.* The following list of examples may be enlarged by placing radicals over the denominators of those in Exercise LXX.

1. $\int \frac{dx}{\sqrt{1-x^2}}$.
2. $\int \frac{dx}{\sqrt{x^2-1}}$.
3. $\int \frac{dx}{\sqrt{4-9x^2}}$.
4. $\int \frac{dx}{\sqrt{4x^2+3}}$.
5. $\int \frac{dx}{\sqrt{9-7x^2}}$.
6. $\int \frac{2\,dx}{\sqrt{5x^2+2}}$.
7. $\int \frac{dx}{\sqrt{2x-x^2}}$.
8. $\int \frac{dx}{\sqrt{x^2+2x}}$.
9. $\int \frac{dx}{\sqrt{7x+4x^2}}$.
10. $\int \frac{dx}{\sqrt{3x-2x^2}}$.
11. $\int \frac{dx}{\sqrt{15+2x-x^2}}$.
12. $\int \frac{dx}{\sqrt{13-12x-x^2}}$.
13. $\int \frac{dx}{\sqrt{16-6x-x^2}}$.
14. $\int \frac{dx}{\sqrt{11-6x+x^2}}$.
15. $\int \frac{dx}{\sqrt{1-6x-9x^2}}$.
16. $\int \frac{dx}{\sqrt{1-4x-4x^2}}$.
17. $\int \frac{2\,dx}{\sqrt{5-4x-3x^2}}$.
18. $\int \frac{6\,dx}{\sqrt{9-8x+7x^2}}$.
19. $\int \frac{e^{2x}dx}{\sqrt{e^{4x}-4}}$.
20. $\int \frac{dx}{e^x\sqrt{1-e^{-2x}}}$.
21. $\int \frac{dx}{x\sqrt{3-\log^2 x}}$.
22. $\int \frac{\cos x/2\,dx}{\sqrt{\sin^2 x/2+5}}$.
23. $\int \frac{\sec^2 x\,dx}{\sqrt{\tan^2 x+9}}$.
24. $\int \frac{e^{\sin x}\cos x\,dx}{\sqrt{e^{2\sin x}+4}}$.

EXERCISE LXXII

Linear over Quadratic. The forms here are of the type $\int (L/Q)\,dx$. Here, as previously, Q represents the general quadratic function and L the general linear function $mx+n$. In case $L\,dx = dQ$ the integral reduces to that of Exercise LXVII. Otherwise the integral must be broken up into two simpler integrals of the type $\int (1/v)\,dv$ and $\int (1/Q)\,dx$. In the case where Q has no middle term the integral $\int \frac{mx+n}{ax^2+c}dx = m\int \frac{x\,dx}{ax^2+c} + n\int \frac{dx}{ax^2+c}$, and both these integrals have been treated in the previous exercises. Hence the only novelty arises in the case $b \neq 0$. In this case the integrand can be broken up by inspection, but the use of undetermined coefficients has many advantages. Thus

$$\int \frac{mx+n}{ax^2+bx+c}\,dx = A\int\frac{2\,ax+b}{ax^2+bx+c}\,dx + B\int\frac{\sqrt{a}\,dx}{ax^2+bx+c}$$
$$= A\log(ax^2+bx+c) + B\int\frac{\sqrt{a}\,dx}{ax^2+bx+c}.$$

This last integral is integrated as in Exercise LXX. A and B are then determined from the identity $mx + n \equiv 2\,A\,ax + Ab + B\sqrt{a}$. Equating coefficients of like powers of x, we have $m = 2\,A$ and $n = Ab + B\sqrt{a}$, which we can solve for A and B.

Illustration:

$$\int\frac{(2\,x+3)\,dx}{9\,x^2+6\,x+4} \equiv A\int\frac{18\,x+6}{9\,x^2+6\,x+4} + B\int\frac{3\,dx}{(3\,x+1)^2+3}$$
$$= A\log(9\,x^2+6\,x+4) + \frac{B}{\sqrt{3}}\arctan\frac{3\,x+1}{\sqrt{3}}.$$

To determine A and B, we have the identity

$$2\,x+3 \equiv 18Ax + 6A + 3B,$$

whence $18A = 2$, $6A + 3B = 3$. Therefore $A = \frac{1}{9}$, $B = \frac{7}{9}$.

Therefore

$$\int\frac{(2\,x+3)\,dx}{9\,x^2+6\,x+4} = \frac{1}{9}\log(9\,x^2+6\,x+4) + \frac{7}{9\sqrt{3}}\arctan\frac{3\,x+1}{\sqrt{3}}.$$

1. $\int\frac{(x+1)\,dx}{x^2+1}$.
2. $\int\frac{(x-2)\,dx}{x^2-1}$.
3. $\int\frac{(2\,x+3)\,dx}{4\,x^2+1}$.
4. $\int\frac{(6\,x-1)\,dx}{1-9\,x^2}$.
5. $\int\frac{(2\,x-3)\,dx}{3\,x^2-2}$.
6. $\int\frac{(5\,x+2)\,dx}{5\,x^2+2}$.
7. $\int\frac{(2+x)\,dx}{11+x^2}$.
8. $\int\frac{(7-3\,x)\,dx}{5-2\,x^2}$.
9. $\int\frac{(x+2)\,dx}{x^2+2\,x}$.
10. $\int\frac{(x+3)\,dx}{6\,x-x^2}$.
11. $\int\frac{(2\,x+3)\,dx}{x^2+3\,x}$.
12. $\int\frac{(6-x)\,dx}{5\,x-x^2}$.
13. $\int\frac{(2\,x+5)\,dx}{x^2+2\,x+5}$.
14. $\int\frac{(x-6)\,dx}{x^2-4\,x-5}$.
15. $\int\frac{(3x-5)\,dx}{x^2+8\,x+20}$.

16. $\int \frac{(5x-3)\,dx}{x^2-6x-7}$.

17. $\int \frac{x\,dx}{x^2-8x+18}$.

18. $\int \frac{x\,dx}{2-6x-x^2}$.

19. $\int \frac{x\,dx}{x^2+x+1}$.

20. $\int \frac{x\,dx}{x^2-x-1}$.

21. $\int \frac{(1-x)\,dx}{4x^2-4x-3}$.

22. $\int \frac{(1-6x)\,dx}{9x^2+6x+5}$.

23. $\int \frac{(3x-2)\,dx}{1-6x-9x^2}$.

24. $\int \frac{(2x+3)\,dx}{9x^2+6x+4}$.

25. $\int \frac{(3x+2)\,dx}{9x^2+3x-2}$.

26. $\int \frac{(10x-3)\,dx}{25x^2-5x+1}$.

27. $\int \frac{(x+2)\,dx}{3x^2+2x-1}$.

28. $\int \frac{(x-2)\,dx}{5x^2+4x+2}$.

29. $\int \frac{(1-x)\,dx}{3x^2-4x+3}$.

30. $\int \frac{(2-x)\,dx}{5x^2-2x+3}$.

31. $\int \frac{(5x+6)\,dx}{7x^2-8x+9}$.

32. $\int \frac{(6x-5)\,dx}{9x^2-6x+7}$.

33. $\int \frac{(11x+2)\,dx}{8-3x+2x^2}$.

34. $\int \frac{(3x-7)\,dx}{4-5x-3x^2}$.

35. $\int \frac{(2x-5)\,dx}{1-2x-6x^2}$.

36. $\int \frac{(x+8)\,dx}{3x^2+5x-12}$.

EXERCISE LXXIII

Linear over the Square Root of a Quadratic. The integrands here are all of the form $L/\sqrt{Q}$. The method used in Exercise LXXII is also applicable to the integration of this form, the two resultant integrals being $\int (1/\sqrt{Q})\,dQ$ * and $\int (1/\sqrt{Q})\,dx$, which are treated in Exercises LXVI and LXXI respectively. Additional examples can be had by placing radicals over the denominators in Exercise LXXII. The breaking up of the integral into two simpler integrals is accomplished precisely as in Exercise LXXII, as shown in the following illustrations. Undetermined coefficients can be used as before.

(a) $$\int \frac{(2x+3)\,dx}{\sqrt{4x^2+1}} = 2\int \frac{x\,dx}{\sqrt{4x^2+1}} + 3\int \frac{dx}{\sqrt{4x^2+1}}$$
$$= \tfrac{1}{2}\sqrt{4x^2+1} + \tfrac{3}{2}\log\left(2x+\sqrt{4x^2+1}\right) + C.$$

* Owing to the frequency with which this particular case of the power form ($n=-1/2$) occurs, it is well that the student should include it in his list of formulas, $\int (1/\sqrt{v})\,dv = 2\sqrt{v} + C$.

(b) $\int \frac{(2x+3)\,dx}{\sqrt{9x^2+6x+4}}$

$$= A\int \frac{(18x+6)\,dx}{\sqrt{9x^2+6x+4}} + B\int \frac{3\,dx}{\sqrt{(3x+1)^2+3}}$$

$$= 2A\sqrt{9x^2+6x+4} + B\log\left[(3x+1)+\sqrt{9x^2+6x+4}\right] + C$$

$$= \tfrac{2}{9}\sqrt{9x^2+6x+4} + \tfrac{7}{9}\log\left[(3x+1)+\sqrt{9x^2+6x+4}\right] + C.$$

A and B are determined as in the illustration in Exercise LXXII, and have the same values, of course.

1. $\int \frac{(2x+1)\,dx}{\sqrt{x^2-1}}$.
2. $\int \frac{(x-1)\,dx}{\sqrt{1-x^2}}$.
3. $\int \frac{(6x+5)\,dx}{\sqrt{9x^2+1}}$.
4. $\int \frac{(1-2x)\,dx}{\sqrt{1-4x^2}}$.
5. $\int \frac{(3x-1)\,dx}{\sqrt{1-3x^2}}$.
6. $\int \frac{(9+x)\,dx}{\sqrt{5+2x^2}}$.
7. $\int \frac{(x+3)\,dx}{\sqrt{x^2+2x}}$.
8. $\int \frac{(x+2)\,dx}{\sqrt{4x-x^2}}$.
9. $\int \frac{x\,dx}{\sqrt{3x+9x^2}}$.
10. $\int \frac{x\,dx}{\sqrt{3x-2x^2}}$.
11. $\int \frac{x\,dx}{\sqrt{27+6x-x^2}}$.
12. $\int \frac{x\,dx}{\sqrt{33-8x-x^2}}$.
13. $\int \frac{(2x+3)\,dx}{\sqrt{14+3x+x^2}}$.
14. $\int \frac{(3x+2)\,dx}{\sqrt{19-5x+x^2}}$.
15. $\int \frac{(4x+5)\,dx}{\sqrt{5-4x-x^2}}$.
16. $\int \frac{(6x+5)\,dx}{\sqrt{6+x-x^2}}$.
17. $\int \frac{(3x-2)\,dx}{\sqrt{4x^2-4x+5}}$.
18. $\int \frac{(5x+2)\,dx}{\sqrt{4x^2+4x+5}}$.
19. $\int \frac{(6x-5)\,dx}{\sqrt{5-9x^2-12x}}$.
20. $\int \frac{(8x-3)\,dx}{\sqrt{12x-4x^2-5}}$.
21. $\int \frac{(x-2)\,dx}{\sqrt{3x^2+6x-2}}$.
22. $\int \frac{(x-1)\,dx}{\sqrt{2x^2+4x+3}}$.
23. $\int \frac{(3x+1)\,dx}{\sqrt{2x^2-8x+3}}$.
24. $\int \frac{(2x+5)\,dx}{\sqrt{3x^2-6x+4}}$.

EXERCISE LXXIV

Integration of Trigonometric Products. The product of various powers of the six fundamental trigonometric functions (of the same angle) can always be reduced to the form $\sin^m\theta\cos^n\theta$, and the corresponding integral $\int \sin^m\theta\cos^n\theta\,d\theta$ can always be solved in two cases.

(a) When one of the exponents m or n is an odd positive integer (no restrictions being placed on the other), it is always possible to take out the factor $\sin\theta d\theta$ (or $\cos\theta d\theta$) which may be written $-d\cos\theta$ (or $d\sin\theta$). The remaining part of the function can then be reduced to an algebraic sum of various powers of $\cos\theta$ (or $\sin\theta$) by using the fundamental relation $\sin^2\theta+\cos^2\theta=1$, and the integral is in each case in the "power form" of Exercise LXVI, where v is $\cos\theta$ (or $\sin\theta$).

(b) When both exponents are even positive integers, the double-angle substitutions

$$\sin^2\theta=\tfrac{1}{2}(1-\cos 2\theta),\quad \cos^2\theta=\tfrac{1}{2}(1+\cos 2\theta),\quad \sin\theta\cos\theta=\tfrac{1}{2}\sin 2\theta,$$

should be used repeatedly until the trigonometric integrals in each case reduce to odd powers of the cosines and sines of even multiples of θ. Integration is then accomplished as in Exercise LXIX or as in (a) above.

This classification should not lead the student to believe that these are the only two integrable cases. See examples 31 and 32 of Exercise LXIX, which are not included under either of these cases but are easily integrated by expressing the *entire* function in terms of either $\sin\theta$ or $\cos\theta$ alone. See also examples 1 and 2 of Exercise LXXV.

1. $\int \sin 2x \cos 2x\, dx.$ **3.** $\int \sin x/3 \cos^2 x/3\, dx.$ **5.** $\int \sin^3 x/3\, dx.$

2. $\int \sin^2 3x \cos 3x\, dx.$ **4.** $\int \sin x/2 \cos^3 x/2\, dx.$ **6.** $\int \cos^3 2x/3\, dx.$

7. $\int (\sin 2x/\cos^3 2x)\, dx.$ **11.** $\int \sin^3 \pi x \sqrt{\cos \pi x}\, dx.$

8. $\int (\cos 5x/\sin^2 5x)\, dx.$ **12.** $\int \cos^3 x/3 \sqrt[3]{\sin x/3}\, dx.$

9. $\int \sin^3 x/2 \cos^2 x/2\, dx.$ **13.** $\int \sin 3x/\sqrt{\cos 3x}\, dx.$

10. $\int \cos^3 x/3 \sin^2 x/3\, dx.$ **14.** $\int \cos 4x/\sqrt[3]{\sin 4x}\, dx.$

15. $\int \sin^5 x/5\, dx.$ **18.** $\int \cot x/\sqrt{\sin x}\, dx.$ **21.** $\int \sin^2 x/3\, dx.$

16. $\int \cos^7 7x\, dx.$ **19.** $\int (\tan^3 x/\sec^2 x)\, dx.$ **22.** $\int \cos^2 x/5\, dx.$

17. $\int \tan x/\sqrt{\sec x}\, dx.$ **20.** $\int (\sec^3 x/\tan^4 x)\, dx.$ **23.** $\int \sin^2 2x \cos^2 2x\, dx.$

24. $\int \sin^2 x/2 \cos^2 x/2 \, dx.$

• 25. $\int \sin^4 x/2 \, dx.$

26. $\int \cos^4 4x \, dx.$

27. $\int \sin^2 x/2 \cos^4 x/2 \, dx.$

28. $\int \cos^2 2x \sin^4 2x \, dx.$

29. $\int \cos^6 (1-x) \, dx.$

30. $\int \sin^6 (2x+1) \, dx.$

31. $\int \sin^4 \theta/4 \cos^4 \theta/4 \, d\theta.$

32. $\int \sin^6 \theta/4 \cos^2 \theta/4 \, d\theta.$

33. $\int (1+\sin 2x)^2 dx.$

34. $\int (1-\cos x/2)^2 dx.$

35. $\int (\sin x + \cos^2 x)^2 dx.$

36. $\int (\sqrt{\sin x} + \cos x)^2 dx.$

37. $\int \sin x \cos^3 x \tan^2 x \, dx.$

38. $\int (\cos^2 x/\csc x \tan^8 x) \, dx.$

39. $\int \sin 2x \sin x \, dx.$

40. $\int \cos 2x \cos x \, dx.$

41. $\int (\sin 2x + \sin x)^2 dx.$

42. $\int (\cos x - \sin 2x)^2 dx.$

43. $\int \left(\sin \frac{x}{2} - 2 \cos x\right)^2 dx.$

EXERCISE LXXV

Integration of Trigonometric Products (*continued*). Trigonometric products are frequently more simply written in one of the following forms:

$$(1) \int \tan^r\theta \sec^s\theta \, d\theta, \qquad (2) \int \cot^r\theta \csc^s\theta \, d\theta.$$

Since these two integrals are analogous, we shall discuss the first one only. There are two methods suggested for attacking the problem: one, to take out the factor $\sec^2\theta \, d\theta$ (that is, $d \tan \theta$) and express the remaining factor or factors of the integrand in terms of $\tan\theta$ alone by using the relation $\sec^2\theta = 1 + \tan^2\theta$; the other, to take out the factor $\sec\theta \tan\theta \, d\theta$ (that is, $d \sec\theta$) and express the remaining factor in terms of $\sec\theta$ alone by using the relation $\tan^2\theta = \sec^2\theta - 1$. The resulting integrals can then be solved by means of the power form.

Following these natural suggestions there are two cases when product (1) above can be integrated in this manner,

(a) when s is an even positive integer (r unrestricted);

(b) when r is an odd positive integer (s unrestricted). The student should compare these with the integrable cases of the preceding exercise.

The method of this exercise does not include the case where r is even and s odd. If such a form is also nonintegrable by the preceding exercise when expressed in terms of sine and cosine, it can be integrated *by parts*, a device explained in Exercise LXXXVII. An example of an integrable case not included above is Exercise LXXIV, 20. It should be remembered, therefore, that a nonintegrable case of the first form of the trigonometric product (Exercise LXXIV) may fall under one of the integrable cases when thrown into the second form (Exercise LXXV), and vice versa.

1. $\int \sin^2 x \cos^{-4} x\, dx.$
2. $\int \sqrt{\sin x/\cos^5 x} \cdot dx.$
3. $\int \tan 2x \sec^2 2x\, dx.$
4. $\int \tan^3 3x\, dx.$
5. $\int \tan^3 3x \sec 3x\, dx.$
6. $\int \tan x/3 \sec^3 x/3\, dx.$
7. $\int \tan^4 4x\, dx.$
8. $\int \sec^4 x/4\, dx.$
9. $\int \sqrt{\tan x} \sec^2 x\, dx.$
10. $\int \sec^4 x/\sqrt{\tan x} \cdot dx.$
11. $\int (\sec^2 3x/\tan^3 3x)\, dx.$
12. $\int (\sec^4 2x/\tan 2x)\, dx.$
13. $\int \cot^3 x \csc^2 x\, dx.$
14. $\int \cot^3 x \sec^2 x\, dx.$
15. $\int \cot^5 x/2\, dx.$
16. $\int \csc^6 x/3\, dx.$
17. $\int \sec^4 x/5 \tan^2 x/5\, dx.$
18. $\int \sec^2 x/3 \tan^4 x/3\, dx.$
19. $\int \cot^3 \theta \csc^5 \theta\, d\theta.$
20. $\int \tan^5 \theta \sec^4 \theta d\theta.$
21. $\int (\tan x/\cos x)^4\, dx.$
22. $\int (\sin x/\cos^3 x)^2\, dx.$
23. $\int 1/(\sin^4 x \cos^2 x)\, dx.$
24. $\int 1/(\sin^2 x \cos^4 x)\, dx.$

EXERCISE LXXVI

Integration by Trigonometric Substitution. The integration of certain special expressions involving $\sqrt{Q}$, that is, a quadratic under the radical sign and no other irrationality, was accomplished in Exercises LXXI and LXXIII. A much wider list of problems can be integrated by the aid of *trigonometric substitutions*, which reduce the integrand to combinations of integral powers of trigonometric functions. The particular substitution to be used in any example depends on the form assumed by Q when written as the sum (or difference) of two squares. The three cases, with their transformations, are as follows:

$$\text{for } \sqrt{a^2 - v^2} \quad \text{set } v = a\sin\theta, \quad dv = a\cos\theta\, d\theta;$$
$$\text{for } \sqrt{a^2 + v^2} \quad \text{set } v = a\tan\theta, \quad dv = a\sec^2\theta\, d\theta;$$
$$\text{for } \sqrt{v^2 - a^2} \quad \text{set } v = a\sec\theta, \quad dv = a\sec\theta\tan\theta\, d\theta.$$

The resulting trigonometric integrals can be solved by the methods of the preceding exercises. The final result should be expressed in terms of the original variable in these exercises, and a right triangle properly lettered is of practical value in this substitution. For the first case above the triangle is shown in the adjoining figure.

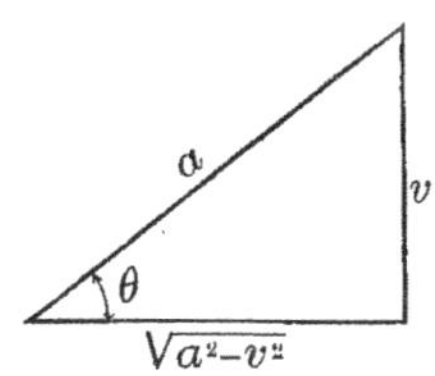

A further substitution which is quite useful in integrating a few special forms involving $\sqrt{Q}$ will be given in Exercise LXXXIV.

1. $\int \frac{dx}{\sqrt{9 - x^2}}$.

2. $\int \frac{dx}{x\sqrt{x^2 - 4}}$.

3. $\int \frac{dx}{x^2\sqrt{x^2 - 3}}$.

4. $\int \frac{dx}{x^2\sqrt{5 - x^2}}$.

5. $\int \frac{dx}{(\sqrt{x^2 + 2})^3}$.

6. $\int \frac{dx}{(\sqrt{2 - x^2})^3}$.

7. $\int \frac{\sqrt{4 - x^2}\,dx}{x^2}$.

8. $\int \frac{\sqrt{x^2 - 16}\,dx}{x}$.

9. $\int \frac{dx}{x\sqrt{x^2 + 9}}$.

10. $\int \frac{dx}{x\sqrt{9 - x^2}}$.

11. $\int \frac{\sqrt{25 - x^2}\,dx}{x}$.

12. $\int \frac{\sqrt{x^2 - 36}\,dx}{x^2}$.

13. $\int \frac{\sqrt{x^2 - 8}\, dx}{x^4}$.

14. $\int \frac{dx}{x^2 \sqrt{x^2 + 6}}$.

15. $\int \frac{dx}{x^4 \sqrt{x^2 - 3}}$.

16. $\int x^3 \sqrt{2 - x^2}\, dx$.

17. $\int \frac{dx}{x^4 \sqrt{x^2 + 4}}$.

18. $\int \frac{\sqrt{4 + x^2}\, dx}{x^6}$.

19. $\int \frac{x^2\, dx}{\sqrt{9 - x^2}}$.

20. $\int \frac{\sqrt{x^2 - 4}\, dx}{x^3}$.

21. $\int \sqrt{9 - x^2}\, dx$.

22. $\int \frac{dx}{x^3 \sqrt{x^2 - 4}}$.

23. $\int \frac{x^3\, dx}{\sqrt{x^2 + 1}}$.

24. $\int \frac{x^3\, dx}{\sqrt{x^2 - 1}}$.

25. $\int x^3 \sqrt{x^2 + 1}\, dx$.

26. $\int x^3 \sqrt{x^2 - 1}\, dx$.

27. $\int \frac{x^4\, dx}{\sqrt{4 - x^2}}$.

CHAPTER XII

ELEMENTARY APPLICATIONS. DEFINITE INTEGRALS

EXERCISE LXXVII

The Constant of Integration. In each of the following examples the student is asked to find the equation of that curve whose slope at any point is the given function of the coördinates and which passes through the assigned particular point. Separate the variables first, then integrate both the form in x and the one in y. The constant, *before evaluation*, may be added to *either* side.

1. $2x$; $(0, 4)$.
2. $2y$; $(\frac{1}{2}, 1)$.
3. xy; $(2, 1)$.
4. x^2/y; $(2, 3)$.
5. $(x-2)/y$; $(1, 1)$.
6. $(x+1)/(2-y)$; $(0, 4)$.
7. $(2-x)/(y+3)$; $(5, 1)$.
8. y^2/x; $(1, 1)$.
9. $x\sqrt{y}$; $(2, 4)$.
10. $y\sqrt{x}$; $(4, 1)$.
11. $y/(1+x^2)$; $(1, 1)$.
12. $y^2/\sqrt{1-x^2}$; $(0, -1)$.
13. $xy/(1+x^2)$; $(0, 1)$.
14. $\sqrt{1-2y}/x^2$; $(1, -4)$.
15. $\sqrt{(1+x)/(1+y)}$; $(3, 8)$.
16. $\sqrt{(1+y)/(1+x)}$; $(8, 3)$.
17. $x\cos^2 y$; $(2, 0)$.
18. $x^2 \sec y$; $(0, \pi/2)$.
19. $y^2 \sin x$; $(\pi/2, 2)$.
20. $(\cot x)/y$; $(\pi/2, 2)$.
21. $[(\cos y)/x]^2$; $(1, 0)$.
22. $[(\cos x)/y]^2$; $(\pi/2, 0)$.
23. $\sqrt{1-4y^2}/x$; $(0, \frac{1}{4})$.
24. $(1-x^2)/(1+y^2)$; $(1, 1)$.
25. $(1+y)/(1-x^2)$; $(0, 0)$.
26. $xy^2/\sqrt{1+2x^2}$; $(0, 0)$.

27. The velocity of a falling body is $ds/dt = v_0 + 32t$. Find the general expression for s if $v_0 = 16$ and $s = 64$ when $t = 1$.

28. The resistance of the air to an automobile, within certain limits of speed, is proportional to the speed. Hence if F is the net force generated by the motor, we have $M\,dv/dt = F - kv$. Express the velocity in terms of t, knowing that $v = 0$ when $t = 0$.

29. Suppose in example 28 that the power is cut off and the speed is reduced by air resistance alone. Express the subsequent speed in terms of t, assuming a velocity v_0 when the power is cut off.

30. Assume in example 28 that the air resistance is proportional to the square of the speed, giving $M\,dv/dt = F - kv^2$, and show that $v = \sqrt{F/k}\tanh\sqrt{kF}t/M$. [Assume $v = 0$ when $t = 0$. $\operatorname{Tanh} x = (e^x - e^{-x})/(e^x + e^{-x})$]. What is the maximum speed attainable?

31. The resistance of air to a parachute may be regarded as proportional to the square of the velocity. If a man weighs 135 lb. and the parachute 25 lb., and the surface of the parachute is 200 sq. ft., with what maximum velocity will he strike the earth jumping from any height? Assume the air resistance to be 2 lb. per square foot when the velocity is 30 ft. per second. If he reaches the ground in 30 sec., with what velocity will he land? (In this case the force downward is Mg, where M is the combined weight in pounds of man and parachute. The resultant force, as in example 28, is $M dv/dt$.

32. By Hooke's law, the tension in a stretched elastic band or spring is proportional to the amount it is stretched beyond its normal length, or $F = -kx$. We can write F, the tension, not only $M dv/dt$ but also $Mv\,dv/dx$. Assuming that the band is stretched a feet when the mass M is attached (that is, $x = a$ when $v = 0$), show that the velocity for any future position of the mass while this law holds is $v = \sqrt{ka^2 - kx^2}/\sqrt{M}$.

33. Take the result of example 32, $dx/dt = \sqrt{k(a^2 - x^2)/M}$, and show that under the same conditions the position of the particle at any time is $x = a\sin\left(\sqrt{k/M}t + \pi/2\right)$. Evaluate your results of examples 32 and 33 for $k = 16$, $M = 4$ lb., $a = 2$ ft., and check by finding the value of v when $x = 0$, in both cases. It is obvious that the value of v in the two cases should be the same.

34. If the band or spring is hanging vertically and therefore the gravitational force is also acting, the force equation is $Mv\,dv/dx = -kx + Mg$. Assuming that $k = 64$, $M = 4$ lb., $g = 32$ ft./sec.^{2}, and that the spring is stretched 6 ft. when released, show that $v = 4\sqrt{-x^2 + 4x + 12}$.

35. Taking the result of example 34, $dx/dt = 4\sqrt{-x^2 + 4x + 12}$, and the same conditions ($x = 6$ when $t = 0$), show that the position, x, at any time, t, is $x = 4\cos 4t + 2$.

36. Any vertical section of the surface of a liquid contained in a rapidly rotating vessel is defined by the law $x\,dx/dy = P^2g/4\pi^2$, where P is the period of rotation in seconds, and the y-axis is vertical with the origin at the lowest point. Show that the surface is a paraboloid of revolution.

Derive its equation if the vessel is making 120 revolutions per minute, taking $g = 32$. [The defining equation may also be written as $dy/dx = \omega^2 x/g$, where ω is the angular velocity.]

37. A belt passes over a pulley of radius a. Assume the law that the pull, T, in the belt at a distance s from the point where the belt leaves the pulley is given by $dT/ds = \mu T/a$. Calculate the general expression for T in the case where the pull at the point where the belt leaves the pulley is 50 lb. Take μ, the coefficient of friction, as 1/4 and the radius of the pulley as 2 ft.

EXERCISE LXXVIII

Orthogonal Trajectories. An orthogonal trajectory of a given system of curves is a curve which cuts each curve of the given system at right angles. Therefore, to find the equation of the orthogonal trajectories we must first find dy/dx from the equation of the given system. Then set dy/dx equal to the negative reciprocal of the derivative just found and integrate. Find the equation of the system of orthogonal trajectories of each of the following systems:

1. $y^2 = 2x + C$. **3.** $xy = C$. **5.** $y^2 + x^2 = C$. **7.** $y^2 + 4x^2 = C$.

2. $y = \frac{1}{3}x^3 + C$. **4.** $y^4 x = C$. **6.** $y^2 - x^2 = C$. **8.** $y^2 - 4x^2 = C$.

9. $y = \frac{1}{2}\log(1 - x^2) + C$. **13.** $y = \tan x + C$. **17.** $x^3 + y^3 = C$.

10. $y = \log C(1 + x^3)$. **14.** $y = \sec x + C$. **18.** $x^2 + y^3 = C$.

11. $y = 1/(1 + x) + C$. **15.** $x^{\frac{1}{2}} + y^{\frac{1}{2}} = C$. **19.** $y^2 = 4Cx$.

12. $y = x^4 + C$. **16.** $x^{\frac{2}{3}} + y^{\frac{2}{3}} = C$. **20.** $y^2 + 2x^2 = C^2$.

EXERCISE LXXIX

The Definite Integral. Evaluation by Direct Integration. The process of evaluating a definite integral consists of first finding the indefinite integral, dropping the constant, as it disappears any way. Evaluate this integral for the upper limit b. Call this value B. Evaluate for the lower limit a, calling this value A. Then *if the function is continuous between the limits*, $B - A$ is the value of the definite integral. Before evaluating a definite integral the student should test the integrand to see whether it becomes infinite (or discontinuous) for any value of x between the limits. If the integrand becomes infinite for a limit, say b, then

$$\int_a^b f(x)\,dx = \underset{h \doteq 0}{\text{limit}} \int_a^{b-h} f(x)\,dx,$$

which may be finite or infinite. Or if it becomes infinite for $x = x_1$, a value between a and b, the integral

$$\int_a^b f(x)\,dx = \underset{h \doteq 0}{\text{limit}} \int_a^{x_1-h} f(x)\,dx + \underset{h \doteq 0}{\text{limit}} \int_{x_1+h}^{b} f(x)\,dx,$$

and if the limit of either of the integrals on the right is ∞, the definite integral in question has no finite values. Examples 10, 16, and 24 are among those requiring examination. Why is example 24 meaningless?

1. $\int_0^1 x^2(1-\sqrt{x})\,dx.$
2. $\int_1^0 (1+\sqrt{x})^2\,dx.$
3. $\int_2^6 (1-x/2)^3\,dx.$
4. $\int_1^2 \frac{dx}{(3-x)^4}.$
5. $\int_3^8 \frac{dx}{\sqrt{1+x}}.$
6. $\int_1^4 \sqrt{5-x}\,dx.$
7. $\int_0^2 \frac{dx}{\sqrt{9-4x}}.$
8. $\int_{-1}^{\frac{4}{3}} \frac{2\,dx}{\sqrt[3]{5-3x}}.$
9. $\int_{-12}^1 \frac{4\,dx}{\sqrt[3]{3-2x}}.$
10. $\int_{-5}^2 \frac{6\,dx}{\sqrt[3]{2-5x}}.$
11. $\int_3^5 \frac{x\,dx}{\sqrt{x^2-1}}.$
12. $\int_5^9 \frac{x\,dx}{\sqrt{x^2+144}}.$
13. $\int_{\frac{\pi}{2}}^{\frac{\pi}{4}} \frac{7\cos 2x\,dx}{\sqrt{9-5\sin 2x}}.$
14. $\int_{\frac{\pi}{6}}^{\frac{\pi}{8}} \frac{4\sin 3x\,dx}{\sqrt{9+5\cos 3x}}.$
15. $\int_3^5 \frac{x\,dx}{x^2-1}.$
16. $\int_0^4 \frac{x\,dx}{(x^2-4)^{\frac{2}{3}}}.$
17. $\int_0^{\frac{\pi}{4}} \frac{\cos 2x\,dx}{1+\sin 2x}.$
18. $\int_0^{\pi} \frac{\sin\frac{x}{2}\,dx}{1+\cos\frac{x}{2}}.$
19. $\int_0^{\sqrt{8}} \frac{dx}{x^2+9}.$
20. $\int_0^{\frac{1}{2}} \frac{8\,dx}{3+4x^2}.$
21. $\int_0^{\infty} \frac{dx}{x^2+4}.$
22. $\int_0^{\infty} \frac{dx}{4x^2+1}.$
23. $\int_{-\sqrt{8}}^{+\sqrt{8}} \frac{7\,dx}{2x^2-25}.$
24. $\int_{-3}^{3} \frac{x\,dx}{\sqrt{x^2-4}}.$
25. $\int_3^6 \frac{3\,dx}{x^2-6x+18}.$
26. $\int_{-1}^{\sqrt{2}-1} \frac{4\,dx}{x^2+2x+3}.$
27. $\int_0^1 \frac{2\,dx}{\sqrt{4-3x^2}}.$
28. $\int_0^{\sqrt{2}} \frac{dx}{\sqrt{2-x^2}}.$
29. $\int_3^4 \frac{dx}{\sqrt{x^2+2x-8}}.$
30. $\int_0^4 \frac{4\,dx}{(x-2)^2}.$
31. $\int_{\frac{1}{3}}^0 \sin \pi x\,dx.$

32. $\int_0^{\frac{\pi}{6}} \cos^2 \frac{x}{2}\, dx.$ 35. $\int_{\frac{\pi}{3}}^{\frac{\pi}{2}} \sin^3 \frac{x}{2} \cos \frac{x}{2}\, dx.$ 38. $\int_{-\infty}^{\log 5} \frac{dx}{e^x + 25 e^{-x}}.$

33. $\int_0^{\frac{\pi}{4}} \sin^3 2x\, dx.$ 36. $\int_{\frac{\pi}{2}}^{\pi} \tan^2 \frac{x}{3} \sec^2 \frac{x}{3}\, dx.$ 39. $\int_0^{\frac{\pi}{3}} \tan^3 x \sec x\, dx.$

34. $\int_{\frac{\pi}{4}}^{\pi} \tan^4 x\, dx.$ 37. $\int_{-\infty}^{\log 3} \frac{dx}{e^x + 9 e^{-x}}.$ 40. $\int_0^{\infty} \frac{8 a^3 dx}{x^2 + 4 a^2}.$

41. $\int_{3/2}^{6} \frac{4\, dx}{2x - 3}.$ 42. $\int_{3/2}^{6} \frac{4\, dx}{\sqrt{2x - 3}}.$

EXERCISE LXXX

Simple Plane Areas. The differential of the area between a curve and the x-axis has been shown to be $y\,dx$. Hence the area between a curve and the x-axis, and between two ordinates $x = a$ and $x = b$, is the definite integral $\int_a^b y\, dx$. If a continuous curve cuts the x-axis twice, we can find the area bounded by the "loop" of the curve and the x-axis. It must be observed that if the lower limit is the smaller of the two numbers a and b, the areas above the x-axis are positive and those below negative. Hence *if an area is partly above and partly below the x-axis, the two parts must be calculated separately.* The subject of areas will be taken up again in later exercises; cf. Exercises XCV, CV, etc.

In each of the examples find the area between the given curve and the x-axis.

1. $y = x^2 - 6x + 5.$ 4. $y = x^3 - 4x^2.$ 7. $y^2 = 4x^2 - x^4$ (total area).

2. $y = 1 - x^2.$ 5. $y = \sin 2x.$

3. $y = x^3 - 4x.$ 6. $y = \cos \frac{1}{2} x.$ 8. $3y = x^2 - 3x - 18.$

Further simple examples may be taken from Exercise XCV, 1.

In the remaining cases find the area between the ordinates indicated in the parentheses.

9. $y = 4/x^2$, (1, 4). 12. $y = (x + 2)/(x^2 + 4x)$, (1, 3).

10. $y = e^{-x/4}$, (0, 4). 13. $y = e^x - e^{-x}$, (0, 2).

11. $y = 4/(x + 1)$, (1, 3). 14. $y = e^{-x/4} + \sin x$, (0, π).

15. A river bends around a meadow, making a curve $y = x - 2x^2$, where the unit is 1 mile and a road is the x-axis. How many acres between the road and the river?

16. A stream has the shape $4y = x^3 - 4x^2 + 4x$, where a road is the x-axis crossing it at the origin. What is the value of land between the river and the road, from the point where the road crosses the stream to the point where it comes to the river bank again, land being worth $60 per acre. The unit is 1 mile.

CHAPTER XIII

SPECIAL METHODS OF INTEGRATION. APPLICATIONS

EXERCISE LXXXI

Integration of Rational Fractions. The integrand is of the form $f(x)/F(x)$, where both $f(x)$ and $F(x)$ are polynomials. The examples in this list cover those cases where the denominator, $F(x)$, contains linear factors, single or repeated, and factors of the second degree, but none repeated. If the denominator consists of or contains a factor of the second degree repeated, the integral is in general more readily obtained by a reduction formula or by a table of integrals, so no examples are here included. The student should observe the following rules carefully: (1) If $f(x)$ is of the same or higher degree than $F(x)$, reduce the integrand to a mixed quantity before attempting integration. (See (a) below.) (2) If the integrand is a proper fraction and the denominator can be factored, it should be broken up into partial fractions by means of undetermined coefficients. The resultant expression is integrated, term by term, by methods already learned, and it is frequently desirable to integrate before calculating the undetermined coefficients. The outline of a few examples should make this clear. Note in (b) and (c) that, whereas $Ax+B$ is the numerator over an unfactorable quadratic, in the case of a linear factor squared, as $(ax+b)^2$, the fraction is broken into $A/(ax+b)+B/(ax+b)^2$. (In the omitted case of quadratics squared in the denominator, the process is similar.)

$$\text{(a)} \int \frac{x^3+7}{x+1}\,dx = \int\left(x^2 - x + 1 + \frac{6}{x+1}\right)dx$$

$$= \tfrac{1}{3}x^3 - \tfrac{1}{2}x^2 + x + 6\log(x+1) + C.$$

$$\text{(b)} \int \frac{(9x+5)\,dx}{x(x+1)(x^2+2x+5)}$$
$$= \int \frac{A\,dx}{x} + \int \frac{B\,dx}{x+1} + \int \frac{Cx+D}{x^2+2x+5}$$
$$= A \log x + B \log (x+1) + \frac{C}{2} \log (x^2+2x+5)$$
$$+ \frac{D-C}{2} \tan^{-1} \frac{x+1}{2}.$$

And by the usual method we get $A = 1$, $B = 1$, $C = -2$, $D = -3$. The student should be clear as to the process here. We are trying to find A, B, C, and D such that the integrand $\frac{9x+5}{x(x+1)(x^2+2x+5)} \equiv \frac{A}{x} + \frac{B}{x+1} + \frac{Cx+D}{x^2+2x+5}$, and this is accomplished by clearing of fractions and equating coefficients. By this method we get the above values of A, B, C, and D.

$$\text{(c)} \int \frac{(20x+17)\,dx}{(2x+1)^2(3x+5)}$$
$$= \int \frac{A\,dx}{2x+1} + \int \frac{B\,dx}{(2x+1)^2} + \int \frac{C\,dx}{3x+5}$$
$$= \frac{A}{2} \log (2x+1) - \frac{B}{2(2x+1)} + \frac{C}{3} \log (3x+5).$$

Next we find $A = 2$, $B = 2$, $C = -3$.

1. $\int \frac{x\,dx}{x-1}$.
2. $\int \frac{(x+2)\,dx}{x+1}$.
3. $\int \frac{(4x-5)\,dx}{3-2x}$.
4. $\int \frac{(2x+3)\,dx}{3x+2}$.
5. $\int \frac{(x^2+1)\,dx}{x-1}$.
6. $\int \frac{(x^3+1)\,dx}{x-1}$.
7. $\int \frac{(x^3-2x)\,dx}{x+1}$.
8. $\int \frac{(x^3-x^2)\,dx}{x+1}$.
9. $\int \frac{(x^3+1)\,dx}{x^2+1}$.
10. $\int \frac{(4x^2+1)\,dx}{x^2+4}$.
11. $\int \frac{x^3\,dx}{x^2+x+1}$.
12. $\int \frac{(x^2+x+1)\,dx}{x^2-x+1}$.
13. $\int \frac{(8x+2)\,dx}{x-x^3}$.
14. $\int \frac{(x+1)\,dx}{x^2+5x+6}$.
15. $\int \frac{(4x^2+8x+1)\,dx}{4x^3-x}$.
16. $\int \frac{(9x^2+12x+1)\,dx}{9x^3-x}$.

17. $\int \frac{(2x^2+5x+6)\,dx}{x^3+5x^2+6x}.$

18. $\int \frac{(2x^2+7x-6)\,dx}{x^3+x^2-2x}.$

19. $\int \frac{(7x^2-x-6)\,dx}{6x^3+13x^2+6x}.$

20. $\int \frac{(10-9x-19x^2)\,dx}{10x^3+29x^2+10x}.$

21. $\int \frac{4x\,dx}{(x+1)(x+2)(x+3)}.$

22. $\int \frac{(x^2+x-3)\,dx}{(x-1)(x-2)(x-3)}.$

23. $\int \frac{(x^2-x+6)\,dx}{x^4-10x^2+9}.$

24. $\int \frac{(x^2-3x-2)\,dx}{x^4-5x^2+4}.$

25. $\int \frac{(2x^2-1)\,dx}{(x^2-2x)(x^2-1)}.$

26. $\int \frac{(7x^2-4)\,dx}{(x^2+x)(x^2-4)}.$

27. $\int \frac{(1+x-3x^2)\,dx}{(x^4-2x^3-x^2+2x)}.$

28. $\int \frac{(2x^3+3x^2-4)\,dx}{x^4+x^3-4x^2-4x}.$

29. $\int \frac{(3-x)\,dx}{x^3+x^2}.$

30. $\int \frac{(x^3-2)\,dx}{x^3-x^2}.$

31. $\int \frac{dx}{2x^3-x^2}.$

32. $\int \frac{(2-7x)\,dx}{2x^2-3x^3}.$

33. $\int \frac{(2x^2+4x+1)\,dx}{x^3+2x^2+x}.$

34. $\int \frac{(3x^2-4x+4)\,dx}{x^3-4x^2+4x}.$

35. $\int \frac{(x^2+6x-1)\,dx}{(x-3)^2(x-1)}.$

36. $\int \frac{(17+3x-2x^2)\,dx}{(x+1)^2(x-3)}.$

37. $\int \frac{(x^2+4x+6)\,dx}{(x+1)^3}.$

38. $\int \frac{(x^2+x+1)\,dx}{(x-1)^3}.$

39. $\int \frac{(2x^2+9)\,dx}{x^3(x-3)}.$

40. $\int \frac{(x+1)\,dx}{x(x-1)^3}.$

41. $\int \frac{(2x^3-2x^2-2x+1)\,dx}{x^4-2x^3+x^2}.$

42. $\int \frac{(3x^3+12x^2+16x+4)\,dx}{x^4+4x^3+4x^2}.$

43. $\int \frac{(2x^2-x+1)\,dx}{(x^2-x)^2}.$

44. $\int \frac{(4x^4-5x-1)\,dx}{4x^4+4x^3+x^2}.$

45. $\int \frac{(2x^3-x^2+1)\,dx}{(x-2)^4}.$

46. $\int \frac{(2x^3-3x^2+4x-5)\,dx}{(x+3)^5}.$

47. $\int \frac{dx}{x^3+x}.$

48. $\int \frac{(1-x)\,dx}{x^3+x}.$

49. $\int \frac{(x+1)^2\,dx}{x^3+x}.$

50. $\int \frac{(x^4-2)\,dx}{x^3+x}.$

51. $\int \frac{(x-2x^2-9)\,dx}{2x^3+3x}.$

52. $\int \frac{(3x-1)\,dx}{(x^2+1)(x-2)}.$

53. $\int \frac{3\,dx}{(x^2+2)(x-1)}$.

54. $\int \frac{(x^2+2)\,dx}{1+x^3}$.

55. $\int \frac{(x^2+2)\,dx}{1-x^3}$.

56. $\int \frac{2\,dx}{x^3+x^2+x+1}$.

57. $\int \frac{(x+4)\,dx}{4-4x+x^2-x^3}$.

58. $\int \frac{(x^3+4)\,dx}{x^4+2x^2}$.

59. $\int \frac{(4-2x)\,dx}{x^4+2x^2}$.

60. $\int \frac{dx}{3x^4+x^2}$.

61. $\int \frac{dx}{x^2-x^4}$.

62. $\int \frac{3x^2\,dx}{x^4+5x^2+4}$.

63. $\int \frac{15\,dx}{4x^4+17x^2+4}$.

64. $\int \frac{(x+1)^3\,dx}{3x^4+10x^2+3}$.

65. $\int \frac{x\,dx}{54x^4+15x^2+1}$.

66. $\int \frac{(x^2-x)\,dx}{x^4+3x^2+2}$.

67. $\int \frac{(x^2+x+1)\,dx}{x^4+3x^2+2}$.

68. $\int \frac{6\,dx}{2x^4-x^2-1}$.

69. $\int \frac{(3x^2-x+2)\,dx}{(x^2+2)(x^2-x-2)}$.

70. $\int \frac{(x^3+x+3)\,dx}{x^4+x^2+1}$.

71. $\int \frac{(x^2+4x+1)\,dx}{x^4+4x^3+4x^2-1}$.

72. $\int_2^3 \frac{(x^2+1)\,dx}{x-1}$. 73. $\int_0^{\sqrt{3}} \frac{2\,dx}{1-x^4}$. 74. $\int_{\frac{1}{4}}^{\frac{1}{2}} \frac{2\,dx}{x-x^3}$. 75. $\int_2^4 \frac{(3-x)\,dx}{x^3+x^2}$.

EXERCISE LXXXII

Integration by Rationalization. The examples in this section are largely restricted to those involving only the simplest form of an irrationality, that is, the integration of expressions containing fractional powers of the general linear function $ax+b$, or $(ax+b)^{m/n}$. Such expressions are readily reduced to rational fractions, which are the subject matter of the previous exercise. The rationalization is accomplished by the substitution of a new variable z, where $z^n = ax+b$, but all answers should be expressed in terms of the original variable. *Do not forget to transform dx before integrating.*

1. $\int \frac{dx}{1+\sqrt{x}}$.

2. $\int \frac{dx}{1-2\sqrt{x}}$.

3. $\int \frac{dx}{(1+x)\sqrt{x}}$.

4. $\int \frac{dx}{(1-x)\sqrt{x}}$.

5. $\int \frac{dx}{x+2\sqrt{x}+5}$.

6. $\int \frac{dx}{x-2\sqrt{x}-3}$.

7. $\int \frac{dx}{x-4\sqrt{x}}$.

8. $\int \frac{dx}{x+6\sqrt{x}}$.

9. $\int \frac{dx}{x(1-x^{\frac{1}{3}})}$.

10. $\int \frac{dx}{x(1 + x^{\frac{1}{4}})}$.

11. $\int \frac{dx}{\sqrt{x}(1 - x^{\frac{1}{4}})}$.

12. $\int \frac{dx}{\sqrt{x}(1 + x^{\frac{1}{6}})}$.

13. $\int_0^1 \frac{dx}{\sqrt{x}(9 + x^{\frac{1}{3}})}$.

14. $\int \frac{dx}{\sqrt{x}(4 - x^{\frac{1}{5}})}$.

15. $\int_0^3 x\sqrt{1 + x}\, dx$.

16. $\int \frac{x\, dx}{\sqrt{1 + x}}$.

17. $\int_3^8 \frac{dx}{x\sqrt{1 + x}}$.

18. $\int \frac{dx}{x\sqrt{1 - x}}$.

19. $\int_{-1}^2 \frac{dx}{(x + 10)(\sqrt{1 + x})}$.

20. $\int \frac{dx}{(x - 3)\sqrt{1 + x}}$.

21. $\int \frac{dx}{x\sqrt{4 - 3x}}$.

22. $\int \frac{dx}{x\sqrt{3x - 4}}$.

23. $\int \frac{\sqrt{2 + x}\, dx}{1 + x}$.

24. $\int \frac{\sqrt{5 - x}\, dx}{x - 1}$.

25. $\int \frac{\sqrt{2x + 3}\, dx}{3x + 2}$.

26. $\int \frac{\sqrt{3x - 2}\, dx}{2x - 3}$.

27. $\int \frac{x\, dx}{(x + 2)\sqrt{1 + x}}$.

28. $\int \left(\frac{2 - \sqrt{x + 1}}{3 - x}\right) dx$.

29. $\int \frac{dx}{(1 + 6x)^{\frac{1}{2}} + (1 + 6x)^{\frac{1}{3}}}$.

30. $\int \frac{dx}{(1 - 12x)^{\frac{3}{4}} - (1 - 12x)^{\frac{2}{3}}}$.

31. $\int \frac{(3x - 10)\, dx}{(x + 2)(2 - x)^{\frac{3}{2}}}$.

32. $\int \frac{(14 - 5x)\, dx}{(x + 8)(1 - x)^{\frac{3}{2}}}$.

33. $\int_0^2 \frac{dx}{(2 + x)(x + 1)^{\frac{3}{2}}}$.

34. $\int \frac{9\, dx}{(x + 8)(x - 1)^{\frac{3}{2}}}$.

35. $\int \frac{2\, dx}{(x^2 + 2x)\sqrt{1 + x}}$.

36. $\int \frac{2\, dx}{(x^2 + 4x)\sqrt{x + 1}}$.

37. $\int \frac{dx}{(x^2 - 1)\sqrt{1 - x}}$.

In the remaining examples, rationalize by the substitution suggested.

38. $\int \frac{4\, dx}{x\sqrt{x^2 - 2x + 3}}$. Let $\sqrt{x^2 - 2x + 3} = z - x$.

39. $\int \frac{4x\, dx}{(x^2 - 2x + 3)^{\frac{3}{2}}}$. Let $\sqrt{x^2 - 2x + 3} = z - x$.

40. $\int \frac{2\, dx}{\sqrt{5x - 6 - x^2}}$. Let $\sqrt{5x - 6 - x^2} = (x - 2)z$.

41. $\int \frac{2x\, dx}{\sqrt{5x - 6 - x^2}}$. Let $\sqrt{5x - 6 - x^2} = (x - 2)z$.

42. $\int x^3 (4 + x^2)^{\frac{1}{4}} dx$. Let $4 + x^2 = z^4$.

43. $\int \frac{(4 + x^2)^{\frac{1}{3}}}{x} dx$. Let $4 + x^2 = z^3$.

44. $\int \frac{(9 + 8x^3)^{\frac{1}{3}}}{x^2} dx$. Let $9 + 8x^3 = x^3 z^3$.

45. $\int \frac{(4 + x^4)^{\frac{1}{2}}}{x^3} dx$. Let $4 + x^4 = x^4 z^2$.

The student will find it possible to rationalize a great many forms by a properly chosen substitution. The list above contains a few of the many which might be classified. The principle of rationalizing the integrand is an important one in direct integration.

EXERCISE LXXXIII

Integration of Rational Trigonometric Functions. Such fractions can be reduced to rational algebraic fractions, the integration of which is practiced in Exercise LXXXI. The transformation is $\tan x/2 = z$, which gives $\sin x = 2z/(1 + z^2)$; $\cos x = (1 - z^2)/(1 + z^2)$; $dx = 2\,dz/(1 + z^2)$. The student can derive the formulas for the remaining four trigonometric functions as needed. All answers should be expressed in terms of the original variable by means of the relation $\tan x/2 = z$. This method should not be used unless the expression cannot be integrated by any of the previous simpler methods, as it is often laborious.

1. $\int \frac{dx}{\sin x + \cos x + 1}$.

2. $\int \frac{dx}{\sin x - \cos x + 1}$.

3. $\int \frac{dx}{\tan x + \sin x}$.

4. $\int \frac{dx}{\cot x + \csc x}$.

5. $\int \frac{dx}{5 + 4\cos x}$.

6. $\int \frac{dx}{4 + 5\cos x}$.

7. $\int \frac{dx}{13 - 5\cos x}$.

8. $\int \frac{dx}{13\cos x - 5}$.

9. $\int \frac{dx}{3 + \cos x}$.

10. $\int \frac{dx}{2\cos x + 1}$.

11. $\int \frac{dx}{2 + \sin x}$.

12. $\int \frac{dx}{1 + 2\sin x}$.

13. $\int \frac{dx}{2\sin x - \cos x + 3}$.

14. $\int \frac{dx}{3\sin x - 4\cos x + 5}$.

15. $\int \frac{\cos x\,dx}{5 - 3\cos x}$.

16. $\int \frac{\sin x\,dx}{5 + 4\sin x}$.

17. $\int \frac{\sin x\,dx}{2 - \sin^2 x}$.

18. $\int \frac{\sin x\,dx}{4 - \sin^2 x}$.

19. $\int \frac{dx}{5\sec x - 4}$.

20. $\int \frac{dx}{4\sec x + 5}$.

21. $\int \frac{(1 + \sin x)\,dx}{\cos x - \sin x}$.

22. $\int \frac{(1 - \sin x)\,dx}{\cos x + \sin x}$.

EXERCISE LXXXIV

Integration by the Reciprocal Substitution. By means of this substitution, $x = 1/z$, $dx = -dz/z^2$, various expressions involving $\sqrt{Q}$ can be reduced to forms similar to those of Exercises LXXI and LXXIII. The student should, however, express all answers in terms of the original variable. In particular this substitution should be associated with the forms given in examples 9–20, that is, (1) $\frac{dx}{x\sqrt{Ax^2 + Bx + C}}$ or (2) $\frac{dx}{x^2\sqrt{Ax^2 + Bx + C}}$. The substitution $x = 1/z$ reduces (1) to $-\frac{dz}{\sqrt{A + Bz + Cz^2}}$ and (2) to $\frac{-z\,dz}{\sqrt{A + Bz + Cz^2}}$, both of which are readily integrable or can be found in the tables.

1. $\int \frac{\sqrt{4 - x^2}\,dx}{x^4}$.

2. $\int \frac{dx}{x^2\sqrt{x^2 + 5}}$.

3. $\int \frac{dx}{x\sqrt{9x^2 - 4}}$.

4. $\int \frac{dx}{x\sqrt{25x^2 - 1}}$.

5. $\int \frac{dx}{x\sqrt{x^2 + 2x}}$.

6. $\int \frac{dx}{x\sqrt{3x^2 - 2x}}$.

7. $\int \frac{\sqrt{x^2 - 3x}\,dx}{x^3}$.

8. $\int \frac{\sqrt{2x + 5x^2}\,dx}{x^3}$.

9. $\int \frac{dx}{x\sqrt{3x^2 - 2x - 1}}$.

10. $\int \frac{dx}{x\sqrt{1 + 4x + 5x^2}}$.

11. $\int \frac{dx}{x\sqrt{4x^2 + 5x + 1}}$.

12. $\int \frac{dx}{x\sqrt{8x^2 + 6x + 1}}$.

13. $\int \frac{dx}{x\sqrt{x^2 + 4x - 4}}$.

14. $\int \frac{dx}{x\sqrt{2x^2 - 6x - 9}}$.

15. $\int \frac{dx}{x\sqrt{16+24x+3x^2}}$.

16. $\int \frac{dx}{x\sqrt{9-24x-5x^2}}$.

17. $\int \frac{dx}{x^2\sqrt{1+2x+3x^2}}$.

18. $\int \frac{dx}{x^2\sqrt{3x^2+2x-1}}$.

19. $\int \frac{dx}{x^2\sqrt{27x^2+6x-1}}$.

20. $\int \frac{dx}{x^2\sqrt{33x^2-8x-1}}$.

21. $\int \frac{dx}{(10x+1)\sqrt{x^2+x}}$.

22. $\int \frac{dx}{(3x-1)\sqrt{x^2+x}}$.

23. $\int \frac{\sqrt{x^2+x}\,dx}{x^4}$.

24. $\int \frac{dx}{x^2\sqrt{x^2+x}}$.

25. $\int \frac{x\,dx}{(\sqrt{x^2+3x})^3}$.

26. $\int \frac{(\sqrt{x^2-2x})^3\,dx}{x^5}$.

EXERCISE LXXXV

The Definite Integral. Change in Limits. In each of the following examples the indefinite integral is found by one of the substitutions of earlier exercises. In such problems, when the substitution is made, the limits of the new variable z (or t), corresponding to the limits originally given for x, should be calculated and used in the evaluation, thus eliminating the labor of expressing the indefinite integral in terms of x. Observe the process in these examples:

(a) $\int_3^8 x\sqrt{1+x}\,dx$, where the substitution is $1+x=z^2$.

Hence, when $x=8$, $z=3$, and when $x=3$, $z=2$. Therefore

$$\int_3^8 x\sqrt{1+x}\,dx = 2\int_2^3 z^2(z^2-1)\,dz = 2\left[\frac{z^5}{5}-\frac{z^3}{3}\right]_2^3 = \frac{1076}{15}.$$

In the next example the substitution is $x = 2\sin t$, so when $x=2$, $t=\pi/2$, and when $x=0$, $t=0$.

(b) $$\int_0^2 x^3\sqrt{4-x^2}\,dx = 32\int_0^{\pi/2}\sin^3 t\cos^2 t\,dt$$
$$= 32\int_0^{\pi/2}(\cos^2 t-\cos^4 t)\sin t\,dt$$
$$= -32\left[\frac{\cos^3 t}{3}-\frac{\cos^5 t}{5}\right]_0^{\pi/2} = \frac{64}{15}.$$

1. $\int_4^9 \frac{\sqrt{x}\,dx}{x-1}$.

2. $\int_1^4 \frac{dx}{x+2\sqrt{x}}$.

3. $\int_1^6 x\sqrt{3+x}\,dx$.

4. $\int_0^4 x\sqrt{1+2x}\,dx$.

5. $\int_3^8 \frac{x\,dx}{\sqrt{1+x}}$.

6. $\int_0^5 \frac{x\,dx}{\sqrt{1+3x}}$.

7. $\int_{-\frac{7}{2}}^1 \frac{x\,dx}{\sqrt{2x+7}}$.

8. $\int_{-\frac{3}{2}}^3 \frac{x\,dx}{\sqrt{2x+3}}$.

9. $\int_{-1}^0 \frac{x^3\,dx}{\sqrt{x+1}}$.

10. $\int_2^5 \frac{x\,dx}{(x-1)^{\frac{3}{2}}}$.

11. $\int_{\frac{1}{2}}^{\frac{1}{3}} \frac{dx}{x\sqrt{2x^2+2x+2}}$.

12. $\int_{\frac{1}{2}}^1 \frac{dx}{x\sqrt{5x^2+4x+1}}$.

13. $\int_{-1}^{-\frac{1}{2}} \frac{dx}{x\sqrt{3x^2-2x-1}}$.

14. $\int_{-2}^{-\frac{1}{5}} \frac{dx}{x\sqrt{5x^2-4x-1}}$.

15. $\int_0^{\frac{1}{2}} \frac{x^2\,dx}{\sqrt{1-x^2}}$.

16. $\int_0^1 \sqrt{4-x^2}\,dx$.

17. $\int_0^\infty \frac{x^3\,dx}{\sqrt{1+x^2}}$.

18. $\int_1^2 \frac{\sqrt{x^2-1}\,dx}{x}$.

19. $\int_2^3 \frac{x^3\,dx}{\sqrt{x^2-4}}$.

20. $\int_0^4 \frac{x^2\,dx}{\sqrt{16-x^2}}$.

21. $\int_{\frac{a}{2}}^a \frac{\sqrt{a^2-x^2}}{x^2}\,dx$.

22. $\int_2^5 \frac{3x-1}{\sqrt{x-1}}\,dx$.

23. $\int_0^a \sqrt{a^2-x^2}\,dx$.

24. $\int_0^1 \frac{x^3\,dx}{\sqrt{4-3x^2}}$.

EXERCISE LXXXVI

Integration by Various Devices. The following examples, which must be integrated by methods other than those used in previous exercises, or at least transformed before those methods can be applied, are inserted to encourage original investigation on the part of the student. The solutions should be accomplished without the aid of series, reduction formulas, or tables.

1. $\int \sin\sqrt{x}\,dx$.

2. $\int \sqrt{1+\sin 2x}\,dx$.

3. $\int \sqrt{\frac{1-x}{1+x}}\,dx$.

4. $\int \sqrt{\frac{x}{1-x^3}}\,dx$.

5. $\int \frac{dx}{e^x+1}$.

6. $\int \frac{dx}{e^x+e^{-x}}$.

7. $\int \frac{dx}{\sqrt{x+3}-\sqrt{x+2}}$.

8. $\int \frac{dx}{x-\sqrt{x^2+1}}$.

9. $\int \frac{dx}{\sqrt{e^x+1}}$.

10. $\int (1-x^{\frac{2}{3}})^{\frac{3}{2}}\,dx$.

11. $\int \frac{\sqrt{1+\log x}\,dx}{x}$.

12. $\int \frac{e^{2x}\,dx}{\sqrt{1+e^x}}$.

13. $\int \frac{dx}{\sqrt{1+\sqrt{1+x}}}$.

14. $\int \sqrt{x^2+2x+2}\,dx$.

15. $\int \frac{x^3}{\sqrt{1+x^2}}\,dx$.

16. $\int \sqrt{\frac{1-x}{1+x}}\cdot\frac{dx}{x}$.

17. $\int \frac{\sin 2t\,dt}{\sqrt{1+\sin t}}$.

18. $\int \frac{\sqrt{x}\,dx}{2(1+x)^{\frac{3}{2}}}$.

The method of undetermined coefficients is a device which can be used very effectively if the form of the result is known. For example, it may be applied to $\int F(x)/\sqrt{Q}\cdot dx$, where $F(x)$ is any polynomial in x. In this case we know

$$\int\frac{a_0x^n + a_1x^{n-1} + a_2x^{n-2}\cdots a_n}{\sqrt{ax^2+bx+c}}$$
$$\equiv (A_0x^{n-1} + A_1x^{n-2} + \cdots A_{n-1})\sqrt{ax^2+bx+c}$$
$$+A_n\int\frac{dx}{\sqrt{ax^2+bx+c}}.$$

To find A_0, A_1, etc., we differentiate both sides of this equation, clear of fractions, and then equate the coefficients of like powers of x on the two sides of the identity. Use this device on the next four examples.

19. $\displaystyle\int\frac{2x^2+5}{\sqrt{x^2+2x+4}}dx.$

20. $\displaystyle\int\frac{6x^3+3x^2-5x+3}{\sqrt{2x^2+6x+5}}dx.$

21. $\displaystyle\int\frac{4x^4-2x^3-7x^2-4x-30}{\sqrt{x^2-4x-5}}dx.$

22. $\displaystyle\int\frac{4x^2+6x+3}{\sqrt{5-4x-x^2}}dx.$

Integrals of the form $\displaystyle\int\frac{dx}{a+b\sin x+c\cos x}$ can be reduced to the form $\displaystyle\int\frac{dx}{a+b\cos x}$ by means of the substitution $b = m\sin\alpha$, $c = m\cos\alpha$, where $m=\sqrt{b^2+c^2}$. The actual form is then $\displaystyle\int\frac{dx}{a+m\cos(x-\alpha)}\cdot$ Apply this to the next two examples.

23. $\displaystyle\int\frac{dx}{6+3\sin x+4\cos x}.$

24. $\displaystyle\int\frac{dx}{10+12\sin x+5\cos x}.$

25. $\displaystyle\int\frac{x^3\,dx}{(1+4x)^{5/2}}.$

26. $\displaystyle\int\frac{dx}{(1+x^2)\sqrt{1-x^2}}.$

27. $\displaystyle\int\frac{dx}{(1-x^2)\sqrt{1+x^2}}.$

28. $\displaystyle\int\frac{(2x+3)\,dx}{(x^2+2x+5)^2}.$

29. $\displaystyle\int_0^r \arcsin\sqrt{\frac{x}{r+x}}\,dx.$

30. $\displaystyle\int(a^{2/3}-x^{2/3})^{2/3}\,dx.$

EXERCISE LXXXVII

Integration by Parts. This device is applied to the integration of functions which may be considered as the products of two simpler functions one of which must always admit of integration. The one which admits of integration is combined with the differential of the variable (dx) and denoted by dv. The other is denoted by u. Then $\int u\,dv = uv - \int v\,du$. Example: $\int x \log x\,dx$. Here there is no doubt that $u = \log x$, $dv = x\,dx$.

$$\therefore \int x \log x\,dx = \tfrac{1}{2}x^2 \log x - \int \tfrac{1}{2}x\,dx = \tfrac{1}{2}x^2 \log x - \tfrac{1}{4}x^2.$$

In case both factors of the product are integrable, then we select as part of dv that one which becomes simpler, or at least no more complicated, by integration. If the product consists of an algebraic factor and a trigonometric or exponential one, u is always the algebraic factor. Sometimes it may be necessary to repeat the process of integration by parts in the second integral until the new integral assumes a standard form, as in examples 7, 8, 33, 34.

Certain types of products are also integrated by assuming first one factor and then the other as the integrable part, thus giving rise to two equations from which we eliminate the second integral, thereby solving for the original integral. This method is regularly employed if one factor is exponential and the other trigonometric, as in examples 37–44; also in certain trigonometric products of functions of different angles, as in examples 9–12.

1. $\int x \sin 2x\,dx.$

2. $\int x \cos 3x\,dx.$

3. $\int x \sec^2 2x\,dx.$

4. $\int x \csc^2 3x\,dx.$

5. $\int x \sin^2 x/2\,dx.$

6. $\int x \cos^2 x/2\,dx.$

7. $\int x^2 \sin 2x\,dx.$

8. $\int x^2 \cos x/2\,dx.$

9. $\int \sin x \cos 3x\,dx.$

10. $\int \sin 3x \cos x\,dx.$

11. $\int \sin x \sin 3x\,dx.$

12. $\int \cos x \cos 3x\,dx.$

13. $\int \arctan(x/3)dx.$

14. $\int \arcsin 2x\,dx.$

15. $\int \arctan(2/x)dx.$

16. $\int \arcsin\sqrt{x/2}\,dx.$

17. $\int \arctan(1-x)\,dx.$

18. $\int x \arcsin(1/x)\,dx.$

19. $\int x^3 \arctan x\,dx.$

20. $\int x^3 \arcsin x\,dx.$

21. $\int \frac{\arctan x\,dx}{x^2}.$

22. $\int \frac{x \arcsin x\,dx}{\sqrt{1-x^2}}.$

23. $\int \frac{\arctan\sqrt{x}\,dx}{x^2}.$

24. $\int \frac{\arcsin\sqrt{x}\,dx}{\sqrt{1-x}}.$

25. $\int \log(1-\sqrt{x})\,dx.$

26. $\int \log\left(\frac{2}{x}+1\right)dx.$

27. $\int \frac{\log(2+x)\,dx}{x^2}.$

28. $\int \frac{\log(1+x^2)\,dx}{x^2}.$

29. $\int \sin x \log(1+\sin x)\,dx.$

30. $\int \sec^2 x \log(\sec x - 1)\,dx.$

31. $\int (1-x)\,e^{(1+2x)}\,dx.$

32. $\int (1+2x)\,e^{(1-x)}\,dx.$

33. $\int x^2 e^{3x}\,dx.$

34. $\int x^3 e^{2x}\,dx.$

35. $\int (e^x + x)^2\,dx.$

36. $\int (2^x + x^2)^2\,dx.$

37. $\int e^{2x} \cos 3x\,dx.$

38. $\int e^{3x} \sin 2x\,dx.$

39. $\int e^{x/2} \sin 2x\,dx.$

40. $\int e^{2x} \cos(x/2)\,dx.$

41. $\int \cos\frac{1}{3}x \cdot e^{-x}\,dx.$

42. $\int e^{-x} \sin 3x\,dx.$

43. $\int e^{x/3} \cos x\,dx.$

44. $\int e^{\frac{x}{2}} \sin x\,dx.$

45. $\int \sec^3 x\,dx.$

46. $\int \csc^3 x\,dx.$

47. $\int \tan^2 x \sec x\,dx.$

48. $\int \sec^5 x\,dx.$

49. $\int [xe^x/(1+x)^2]\,dx.$

50. $\int (x^2/e^x)\,dx.$

51. $\int_{1/2}^{1} \log(1+4x^2)\,dx.$

52. $\int \log(a+bx^2)\,dx.$

53. $\int_1^2 \log(4x^2-1)\,dx.$

54. $\int_0^{\frac{\pi}{4}} \frac{2x+\sin 2x}{\cos^2 x}\,dx.$

55. $\int_0^4 x \log\sqrt{9+x^2}\,dx.$

56. $\int_0^{\frac{\pi}{2}} e^{-\frac{x}{2}} \sin 2x\,dx.$

EXERCISE LXXXVIII

Miscellaneous Examples. Review. The following list of examples is inserted here for the purpose of reviewing the various methods of integration which precede, the determination of the method in each example being left to the student. When an example admits of more than one method of solution, the student should in each case use the shortest and least cumbersome one. The integrals may be evaluated between limits if desired. A few limits are suggested.

1. $\int \frac{\arc\sin x\,dx}{\sqrt{1-x^2}}$.
2. $\int \frac{dx}{x\sqrt{2x^2+3x+1}}$.
3. $\int \frac{(5x+2)\,dx}{4x^2+4x+5}$.
4. $\int \frac{(5x+2)\,dx}{\sqrt{4x^2+4x+5}}$.
5. $\int \frac{\sin^3 x\,dx}{\cos^5 x}$.
6. $\int (e^x - e^{-2x})^2\,dx$.
7. $\int \sin^4\frac{x}{4}\cos^2\frac{x}{4}\,dx$.
8. $\int \frac{11\,dx}{\sqrt{3x^2+3x+1}}$.
9. $\int x^2 \sin\frac{x}{3}\,dx$.
10. $\int_5^8 \frac{x\,dx}{\sqrt{x-4}}$.
11. $\int \frac{(x-1)\,dx}{x^2(x^2-2)}$.
12. $\int \frac{13\,dx}{1+x+x^2}$.
13. $\int \frac{(2x+5)\,dx}{\sqrt{x^2+5x}}$.
14. $\int \sqrt{x}(1+x^2)^2\,dx$.
15. $\int \frac{dx}{(x^2+4)^2}$.
16. $\int \sin^6\frac{x}{4}\,dx$.
17. $\int_{-2}^{0} \frac{dx}{x^2+4x+8}$.
18. $\int \frac{dx}{e^x - 16e^{-x}}$.
19. $\int x^2 \arc\tan x\,dx$.
20. $\int \frac{3\,dx}{x\sqrt{4-\log^2 x}}$.
21. $\int \frac{\sec^3 x\,dx}{\tan^4 x}$.
22. $\int_0^1 \frac{dx}{9x^2-6x+1}$.
23. $\int \frac{81\,dx}{x^2(x^2-9)^{\frac{3}{2}}}$.
24. $\int \frac{dx}{x\sqrt{6x^2-5x+1}}$.
25. $\int (1+x)\cos\sqrt{x}\,dx$.
26. $\int \frac{\sqrt{1+x^3}\,dx}{x}$.
27. $\int \frac{(x^2-x)\,dx}{x^2+4}$.
28. $\int \frac{(x-\sin 2x)\,dx}{x^2+\cos 2x}$.
29. $\int \frac{dx}{x\sqrt{1-x^2}}$.
30. $\int x^4(\log x)^2\,dx$.
31. $\int \frac{\sqrt{\arc\tan x}\,dx}{1+x^2}$.
32. $\int \frac{(7x-2)\,dx}{\sqrt{7-2x^2}}$.
33. $\int \frac{(x^2+14)\,dx}{x^4-17x^2+16}$.
34. $\int \sin 2x \cdot e^{\sin^2 x}\,dx$.
35. $\int x\sqrt{1-x^2}\sin^{-1}x\,dx$.
36. $\int \frac{dx}{x^3\sqrt{x^2-4}}$.
37. $\int \frac{(x+1)\,dx}{(x-2)^{\frac{5}{3}}}$.

38. $\int \cos x \sqrt{1+\sin x}\, dx.$

39. $\int x \log(x^2+1)\, dx.$

40. $\int \frac{(4x+5)\, dx}{\sqrt{5-4x-x^2}}.$

41. $\int \frac{\sqrt{x^2+4}\, dx}{x^4}.$

42. $\int \frac{dx}{x(\sqrt{1+x}-2)}$

43. $\int \frac{(4x^2+5)\, dx}{x^3+4x^2+5x}$

44. $\int \sqrt{\frac{\arcsin x\, dx}{1-x^2}}.$

45. $\int \frac{(x+\sin 3x)\, dx}{3x^2-2\cos 3x}.$

46. $\int \frac{(1+\sin x)\, dx}{\sin x\,(2+\cos x)}.$

47. $\int \frac{(x+1)\, dx}{\sqrt{2-x-x^2}}.$

48. $\int \frac{(3-x)^3\, dx}{\sqrt{x}}.$

49. $\int \frac{(x+1)\, dx}{(x-2)^{\frac{5}{3}}}.$

50. $\int \sin^4 \frac{x}{3}\, dx.$

51. $\int (x^2/\sqrt{x^3+1})\, dx.$

52. $\int \frac{(x+1)\, dx}{x^2(x^2+2)}.$

53. $\int_0^1 x^2 \sqrt{1-x^2}\, dx.$

54. $\int e^{\frac{x}{3}} \cos \frac{x}{2}\, dx.$

55. $\int \frac{x^3\, dx}{x+3}.$

56. $\int_1^2 \frac{\sin(1/x)\, dx}{x^2}.$

57. $\int \frac{x \log x\, dx}{\sqrt{1-x^2}}.$

58. $\int \frac{(\sqrt{x}+1)\, dx}{\sqrt{x}\,(x+1)}.$

59. $\int \frac{(2-x^2)\, dx}{2+x^2}.$

60. $\int_0^3 \frac{x\, dx}{\sqrt{4-x}}.$

61. $\int \cos^2 \frac{x}{2}\, dx.$

62. $\int \frac{(x^3+x^2+x-1)\, dx}{x^2(x^2+1)}.$

63. $\int \frac{x\, dx}{x^2-5x+6}.$

64. $\int x^2 \arcsin x\, dx.$

65. $\int \frac{(6+2x^2-3x^3)\, dx}{x^4+3x^2}.$

66. $\int \frac{(x+1)\, dx}{(x+2)^2(x+3)}.$

67. $\int \frac{x^2 \arctan x\, dx}{1+x^2}.$

68. $\int \frac{x^2\, dx}{x^4+x^2-2}.$

69. $\int \frac{(4x-1)\, dx}{\sqrt{1-9x^2}}.$

70. $\int \frac{4\, dx}{x^2\sqrt{1+2x-3x^2}}.$

71. $\int 2x \arctan x\, dx.$

72. $\int a^x b^{2x} e^{3x}\, dx.$

73. $\int \frac{(2x^3-7x^2+1)\, dx}{2x-1}.$

74. $\int \frac{(2x^3+6x-3)\, dx}{1-2x}.$

75. $\int \tan^5 2x\, dx.$

76. $\int \frac{(x^3-3x^2-5x+5)\, dx}{x^2+2x-3}.$

77. $\int \left(\frac{\sin\theta}{\cos^2\theta}+\frac{\cos\theta}{\sin^2\theta}\right) d\theta.$

78. $\int \frac{(1-\cos x)^2}{\sin x}\, dx.$

79. $\int \frac{(1+\sin 2x)\, dx}{\cos^2 2x}.$

80. $\int \sin^2 \tfrac{1}{2}x \cos^2 \tfrac{1}{2}x\, dx.$

81. $\int \arccos x\, dx.$

82. $\int x^2 \log 2x\, dx.$

83. $\int x^2 a^x\, dx.$

84. $\int x^4 (\log x)^2\, dx.$

EXERCISE LXXXIX

Integration by the Use of Series. When a function cannot be integrated by any of the means thus far considered, it is frequently developed into a power series which is integrated term by term. We shall see later that this method is valuable when the resultant series converges. The student should recall the binomial theorem as well as the expansions for the exponential and the trigonometric functions. At least three terms should be obtained in each of the following examples, and, if possible, the *n*th term. (See Exercise LVIII.)

1. $\int\sqrt{1+x^3}\,dx.$	**8.** $\int\sqrt{x}/(1+x)^{\frac{1}{3}}\,dx.$	**15.** $\int(\sin x)/x\,dx.$
2. $\int\sqrt[3]{1+x^2}\,dx.$	**9.** $\int x^{\frac{1}{3}}\sqrt{1-x^2}\,dx.$	**16.** $\int(\cos x^2)/\sqrt{x}\,dx.$
3. $\int\sqrt{1-x^4}\,dx.$	**10.** $\int\sqrt[3]{1-x^3}/\sqrt{x}\,dx.$	**17.** $\int\sqrt{\cos x}\,dx.$
4. $\int\sqrt[3]{1-x^3}\,dx.$	**11.** $\int e^{-x^2}dx.$	**18.** $\int\sqrt{\sec x}\,dx.$
5. $\int 1/\sqrt{1+x^3}\,dx.$	**12.** $\int\sqrt{x}\,e^x\,dx.$	**19.** $\int\sqrt{\sin x}\,dx.$
6. $\int 1/(1+x^2)^{\frac{2}{3}}\,dx.$	**13.** $\int\sin 1/x\,dx.$	**20.** $\int\sqrt{\arctan x}\,dx.$
7. $\int x^{\frac{1}{3}}\sqrt{1+x}\,dx.$	**14.** $\int\cos\sqrt{x}\,dx.$	**21.** $\int\log(x+\sqrt{x^2+4})\cdot dx.$

22. Expand θ in powers of $\tan\theta$. Such expansions are readily obtained by the aid of integration. Proceed as follows. Expand $1/(1+x^2)$ in a power series in x. Integrate both sides, and then let $x=\tan\theta$.

23. Expand θ as a power series in $\sin\theta$. (Start with $1/\sqrt{1-x^2}$.)

EXERCISE XC

Examples for Comparison. The following *groups* of examples are introduced as a means of reviewing and comparing the standard forms and devices for integration used in the previous exercises. The functions in each group bear a certain resemblance to each other, but are integrated by methods essentially different. The exercise is intended not so much for use in actual integration as for a test in discrimination.

1. (a) $\int e^x\,dx$; (b) $\int x^e\,dx$.

2. (a) $\int e^{1/x}/x \cdot dx$; (b) $\int e^{1/x}/x^2 \cdot dx$.

3. (a) $\int xe^{x^2}\,dx$; (b) $\int x^2e^x\,dx$.

4. (a) $\int \sin\sqrt{x}/\sqrt{x}\cdot dx$; (b) $\int \sin x/x \cdot dx$.

5. (a) $\int x\sqrt{1+x}\,dx$; (b) $\int \sqrt{x}\,(1+x)\,dx$.

6. (a) $\int (1-x)/\sqrt{x}\cdot dx$; (b) $\int \sqrt{x}/(1-x)\cdot dx$.

7. (a) $\int x/\sqrt{x^2-4}\cdot dx$; (b) $\int \sqrt{x^2-4}/x \cdot dx$.

8. (a) $\int x\sqrt{9-x^2}\,dx$; (b) $\int 1/x\sqrt{9-x^2}\cdot dx$.

9. (a) $\int e^x \sin e^x\,dx$; (b) $\int x \sin x\,dx$.

10. (a) $\int \tan x \sec^2 x\,dx$; (b) $\tan^2 x \sec x\,dx$.

11. (a) $\int (x^2+2x+3)\,dx$; (b) $\int 1/(x^2+2x+3)\cdot dx$.

12. (a) $\int x/(x^2+1)\cdot dx$; (b) $(x^2+1)/x\cdot dx$.

13. (a) $\int e^{2x}/(e^x+1)\,dx$; (b) $\int e^x/(e^{2x}+1)\cdot dx$.

14. (a) $\int 1/(1+x)\sqrt{x^2+x}\cdot dx$; (b) $\int 1/(x^2+x)\sqrt{1+x}\cdot dx$.

15. (a) $\int \sin^3 x\,dx$; (b) $\int \tan^3 x\,dx$; (c) $\int \sec^3 x\,dx$.

16. (a) $\int \sin^4 x\,dx$; (b) $\int \tan^4 x\,dx$; (c) $\int \sec^4 x\,dx$.

17. (a) $\int 1/(x\log x)\cdot dx$; (b) $\int (\log x)/x\cdot dx$; (c) $\int x\log x\,dx$.

18. (a) $1/\sqrt{1+x^3}\cdot dx$; (b) $\int x^2/\sqrt{1+x^3}\cdot dx$; (c) $\sqrt{x}/\sqrt{1+x^3}\cdot dx$.

19. (a) $\int dx/(x^4-1)$; (b) $\int x/(x^4-1)\cdot dx$; (c) $x^2/(x^4-1)\cdot dx$;
(d) $x^3/(x^4-1)\cdot dx$.

20. (a) $\int 1/(1-\cos\theta)\cdot d\theta$; (b) $\int 1/(2-\cos\theta)\cdot d\theta$;

(c) $\int \sin\theta/(2-\cos\theta)\cdot d\theta$; (d) $\int \cos\theta/(2-\cos\theta)\cdot d\theta$.

EXERCISE XCI

Integration by Reduction Formulas. The following formulas are obtained by applying integration by parts to the functions in question. They are included in one form or another in every table of integrals. As given below, m is any integer, positive or negative, n is a positive integer, and p is any integer or rational fraction.

(1) $\int x^m(a+bx^n)^p dx$

$$=\frac{x^{m-n+1}(a+bx^n)^{p+1}}{(np+m+1)b}-\frac{(m-n+1)a}{(np+m+1)b}\int x^{m-n}(a+bx^n)^p dx.$$

(2) $\int x^m(a+bx^n)^p dx$

$$=\frac{x^{m+1}(a+bx^n)^p}{(np+m+1)}+\frac{anp}{np+m+1}\int x^m(a+bx^n)^{p-1}dx.$$

(3) $\int x^m(a+bx^n)^p dx$

$$=\frac{x^{m+1}(a+bx^n)^{p+1}}{(m+1)a}-\frac{(np+n+m+1)b}{(m+1)a}\int x^{m+n}(a+bx^n)^p dx.$$

(4) $\int x^m(a+bx^n)^p dx$

$$=\frac{-x^{m+1}(a+bx^n)^{p+1}}{n(p+1)a}+\frac{np+n+m+1}{n(p+1)a}\int x^m(a+bx^n)^{p+1}dx.$$

The student should observe just how each of these formulas simplifies the integral, as well as the restriction placed upon the original integrand before one of these formulas can be applied. The collected results follow:

(1) diminishes m by n if $np+m+1\neq 0$;
(2) diminishes p by 1 if $np+m+1\neq 0$;
(3) increases m by n if $m+1\neq 0$;
(4) increases p by 1 if $p+1\neq 0$.

In the cases excepted above, the quantities can be integrated at once by more elementary methods.

In each of the following examples the student should note carefully the numerical equivalent of m, n, and p and should ascertain which formula or formulas to use so as to reduce the same to a standard integral. This exercise may be made an oral one throughout. If the reduction formulas fail, try some other method.

1. $\int x^5\sqrt{1+x^3}\,dx.$
2. $\int \frac{dx}{x\sqrt{1-x^3}}.$
3. $\int \frac{x\,dx}{(1-x^4)^{\frac{3}{2}}}.$
4. $\int \frac{x^5\,dx}{\sqrt{x^4+4}}.$
5. $\int \frac{\sqrt{1+x^4}\,dx}{x}.$
6. $\int \frac{x^8\,dx}{(1+x^3)^{\frac{1}{3}}}.$
7. $\int x^5\sqrt{1-x^4}\,dx.$
8. $\int x\sqrt{1+x^4}\,dx.$
9. $\int \frac{(1+x^5)^{\frac{1}{4}}dx}{x^6}.$
10. $\int \frac{x^8\,dx}{\sqrt{1+x^6}}.$
11. $\int \frac{x^{11}\,dx}{\sqrt{1+x^6}}.$
12. $\int \frac{\sqrt{1+x^6}\,dx}{x^4}.$
13. $\int x^2(1-x^6)^{\frac{3}{2}}\,dx.$
14. $\int \frac{x^9\,dx}{(1+x^5)^3}.$
15. $\int \frac{(1+x^7)^3\,dx}{x}.$
16. $\int \frac{(1+x^8)^{\frac{3}{2}}\,dx}{x}.$
17. $\int x^3(1+x^8)^{\frac{3}{2}}\,dx.$
18. $\int \frac{\sqrt{1+x^3}\,dx}{x^{13}}.$
19. $\int \frac{x^3\,dx}{(1+x^8)^{\frac{5}{2}}}.$
20. $\int \frac{x^{15}\,dx}{(1+x^8)^{\frac{2}{5}}}.$
21. $\int \frac{x^5\,dx}{(9-x^3)^{\frac{5}{2}}}.$
22. $\int \frac{(a^3-x^3)^{\frac{1}{3}}}{x}dx.$
23. $\int \frac{dx}{x(a^4-x^4)^{\frac{2}{3}}}.$
24. $\int \frac{\sqrt{4-x^2}}{x^4}dx.$
25. $\int \frac{dx}{x^3(4-x^4)^{\frac{3}{2}}}.$
26. $\int \frac{x^{\frac{7}{2}}\,dx}{\sqrt{4+x^3}}.$

EXERCISE XCII

Integration by the Aid of Tables. Most of the examples of the following list are integrable by one or more of the methods already studied. They are designed, however, to be integrated here by the aid of an abbreviated table of integrals, such as is contained in many textbooks. They should be sufficient to acquaint the student with the range and content of his table, and to aid him in becoming accurate in its use.

1. $\int \frac{dx}{x\sqrt{1-3x}}$.

2. $\int \sin^2 \frac{x}{4}\, dx$.

3. $\int e^{\frac{x}{2}} \sin 3x\, dx$.

4. $\int e^{2x} \cos 3x\, dx$.

5. $\int \frac{dx}{\sqrt{x^2+2x-3}}$.

6. $\int \frac{dx}{\sqrt{x^2+2x+5}}$.

7. $\int \frac{dx}{\sqrt{3-2x-x^2}}$.

8. $\int \frac{dx}{\sqrt{3-4x-4x^2}}$.

9. $\int \frac{x\,dx}{(4-5x)^2}$.

10. $\int \sqrt{9-4x^2}\, dx$.

11. $\int \frac{x\,dx}{\sqrt{3+6x-9x^2}}$.

12. $\int \frac{dx}{x\sqrt{a^2+x^2}}$.

13. $\int x^4 \log x\, dx$.

14. $\int x^2 \cos 2x\, dx$.

15. $\int \frac{ax^4\, dx}{(b-cx^5)^2}$.

16. $\int \frac{2x+3}{\sqrt{x^2-a^2}}\, dx$.

17. $\int \frac{dx}{x\sqrt{x+1}}$.

18. $\int x \sin 2x\, dx$.

19. $\int \frac{\sqrt{1-x}}{x}\, dx$.

20. $\int \frac{dx}{(x-1)\sqrt{x+2}}$.

21. $\int \frac{dx}{3+\cos x}$.

22. $\int \frac{dx}{2+\sin x}$.

23. $\int \frac{(x-1)\,dx}{\sqrt{x+1}}$.

24. $\int \sqrt{1-2x-3x^2}\, dx$.

25. $\int \frac{(x+2)\,dx}{\sqrt{x^2+9}}$.

26. $\int \frac{dx}{(x+2)\sqrt{x+3}}$.

27. $\int \frac{dx}{x^2\sqrt{9x^2-4}}$.

28. $\int \frac{dx}{\sqrt{4-9x^2}}$.

29. $\int \frac{\cos^4 x\, dx}{\sin^3 x}$.

30. $\int \frac{dx}{x\sqrt{9+x^2}}$.

31. $\int x^2 \log^2 x\, dx$.

32. $\int \frac{\cos x\, dx}{\sin x \log \sin x}$.

33. $\int \frac{dx}{x^2\sqrt{1+3x}}$.

34. $\int x^3 (1+x^2)^{\frac{3}{2}}\, dx$.

35. $\int \frac{x^2\, dx}{e^{2x}}$.

36. $\int \frac{dx}{3+\cos 2x}$.

37. $\int \frac{(4y-5)\,dy}{\sqrt{y^2+2y+5}}$.

38. $\int \frac{(2x+5)\,dx}{\sqrt{x^2+4x+3}}$.

39. $\int \sin^2 x \cos^3 x\, dx$.

40. $\int \frac{\sin^5 x}{\cos^2 x}\, dx$.

41. $\int \frac{dx}{\sqrt{4x-x^2}}$.

42. $\int \frac{e^x\, dx}{\sqrt{3+e^x-e^{2x}}}$.

43. $\int x^2 \sin \frac{x}{2}\, dx$.

44. $\int \frac{x\,dx}{3+2x^2-x^4}$.

45. $\int \frac{x\,dx}{\sqrt{3+2x^2-x^4}}$.

46. $\int \frac{\cos^3 t\, dt}{\sqrt{\sin t}}$.

47. $\int \frac{x^{\frac{1}{2}}\, dx}{\sqrt{2-4x^3}}$.

48. $\int \sqrt{\frac{4+x^2}{3+x^2}} \cdot x\, dx$.

49. $\int \sqrt{\frac{1+\sin x}{2-\sin x}} \cdot \cos x\, dx$.

50. $\int \frac{\sqrt{2ax^2-x^4}}{x^5}\, dx$.

51. $\int \frac{e^{2x}\, dx}{(4e^x-e^{2x})^{\frac{3}{2}}}$.

52. $\int \frac{\sin 2t\, dt}{(2+\sin t)^2}$.

53. $\int e^{-2x}\sqrt{3-2e^{-x}}\, dx$.

54. $\int \frac{\sqrt{2 + x^{3/2}}}{x} dx.$

55. $\int \frac{2x^3 dx}{\sqrt{1 + 4x^2 - 4x^4}}.$

56. $\int \frac{e^{2x} dx}{\sqrt{1 + 2e^x - e^{2x}}}.$

57. $\int \sin 3t \cos 2t \, dt.$

58. $\int \frac{dx}{x^2(1 + x^2)^{3/2}}.$

59. $\int \frac{x^7 dx}{(5 - x^4)^{9/2}}.$

60. $\int \frac{x^3 dx}{(x^2 - 4)^{3/2}}.$

61. $\int x^{7/2}\sqrt{x^3 - 4}\, dx.$

62. $\int x^5 (4 + x^2)^{\frac{1}{3}} dx.$

63. $\int \sin^2 2x \cos^4 2x \, dx.$

64. $\int \frac{5\, dx}{(2x + 1)(5 + 4x)^2}.$

65. $\int \sqrt{4e^x - 1}\, e^{-2x} dx.$

66. $\frac{\cos x \, dx}{\sqrt{-\sin^2 x + 5 \sin x - 6}}.$

EXERCISE XCIII

The Definite Integral. Evaluation by the Use of Tables. Although the student should try to retain the formulas and simpler methods of the preceding exercises, nevertheless, after he has had a thorough drill in formal integration, he should make use of a table of integrals in evaluating the more complicated forms which occur in the applications of the calculus. Skill in using the tables is much to be desired. The examples of this exercise are introduced for practice in such use, no formal integration being required, the evaluation being accomplished by substituting the limits in the indefinite integral in the table.

1. $\int_0^3 (x^2 + 16)^{\frac{3}{2}} dx.$

2. $\int_0^3 x\sqrt{1 + x}\, dx.$

3. $\int_0^1 \sqrt{4x - x^2}\, dx.$

4. $\int_0^4 \sqrt{\frac{5 + x}{x}}\, dx.$

5. $\int_0^4 \sqrt{x^2 + 9}\, dx.$

6. $\int_0^4 y\sqrt{4 - y}\, dy.$

7. $\int_0^{\frac{\pi}{6}} \frac{dx}{\cos^3 x}.$

8. $\int_0^{\frac{\pi}{2}} \frac{dx}{2 + \cos x}.$

9. $\int_1^2 \frac{\sqrt{10x - x^2}\, dx}{x^3}.$

10. $\int_3^4 \frac{dx}{x\sqrt{25 - x^2}}.$

11. $\int_5^{12} \frac{dx}{x\sqrt{4 + x}}.$

12. $\int_5^{12} \frac{\sqrt{4 + x}\, dx}{x}.$

13. $\int_2^3 \frac{dx}{x^2(6 + 3x)}.$

14. $\int_1^2 \frac{dx}{x(4 + 3x^2)}.$

15. $\int_1^2 \frac{\sqrt{x^2 + 5}\, dx}{x}.$

16. $\int_3^5 \frac{\sqrt{x^2 - 1}\, dx}{x^2}.$

17. $\int_0^\pi e^x \sin x \, dx.$

18. $\int_0^1 x e^x dx.$

19. $\int_1^{\sqrt{3}} \frac{x^2 dx}{\sqrt{(4 - x^2)^3}}.$

20. $\int_0^3 \frac{x^2 dx}{\sqrt{(16 + x^2)^3}}.$

EXERCISE XCIV

Applied Problems Involving Definite Integrals. The object of this exercise is to illustrate some types of definite integrals which arise in mechanics, physics, chemistry, etc. In these problems the integral may be regarded as a reversal of differentiation or of rates. The important idea of the definite integral as a summation is illustrated in later exercises. Remember that the variables must be separated before integration.

1. Given the gas equation $pv = k$ and the differential of work $dW = pdv$. Calculate the work done if the volume changes from v_1 to v_2.

2. Given Van der Waals' equation for a gas $(p + a/v^2)(v - b) = k$. Calculate the work done if the volume changes from v_1 to v_2.

3. The formula for the velocity, ds/dt, of a freely falling body is $ds/dt = v_0 + gt$. Taking $g = 32$ ft./sec.^{2}, and $v_0 = 16$ ft./sec., (a) calculate the distance s traversed in the fifth second (*Hint.* The limits of t are 4 and 5); (b) calculate the total distance and the average velocity during the next 5 seconds.

4. The angular velocity, $d\theta/dt$, of a wheel is $\omega_0 + kt$. Assume $\omega_0 = 4\pi$ radians per second, and $k = \pi/6$ rads./sec.^{2} Calculate the number of revolutions the wheel will make during the second minute.

NOTE. In the formula, t is the time in seconds, θ is the angular distance in radians, and ω_0 the initial angular velocity in radians per second.

5. A particle moves on a path along which the slope, dy/dx, is always equal to x^2. How much does y change when x changes from 3 to 6?

6. The formula for the specific heat of iron for ordinary temperatures up to 200° C. is $s = .1053 + .000142\,t$, where t is temperature. Calculate the amount of heat, Q, required to raise 100 kg. of iron from 50° to 100° C., given $dQ/dt = Ms$, where M is the amount of iron in kilograms, and Q is in large calories.

7. A boy starts a slide on ice with a speed of 10 mi. per hour. He stops in $3\frac{2}{3}$ sec., having covered $26\frac{8}{9}$ ft. If his speed at any time is given by $ds/dt = -32\,\mu t + v_0$, where v_0 is his initial speed in feet per second and μ is the coefficient of friction between the boy's shoes and the ice, calculate μ.

8. By Joule's law the quantity of heat, Q, generated in a conductor of constant resistance R is given by the equation $dQ = kRi^2dt$, where i is the current. If i is periodically variable of period P, according to the law $i = i_0 \sin 2\pi t/P$, where i_0 is the maximum value of i, calculate the heat generated in a period (k has a value of nearly $\frac{1}{4}$ if Q is in calories).

9. The work done by an electric current is given by $W = Eit$, where E = electromotive force, i = current, t = time. In an alternating current $E = E_0 \sin(kt + \phi)$ and $i = i_0 \sin(kt + \theta)$, where E_0 and i_0 are maximum values and ϕ and θ are the phase angles (constant). (a) Calculate the work done in a period. (b) Show it is a maximum when $\phi = \theta$.

Hint. $dW = Eidt$.

10. The theorem of Torricelli states that if the water in a vessel is flowing out through a hole in the bottom of the vessel, the velocity of the water as it leaves the orificé is exactly the same as that of a body which has fallen freely under gravity through a distance equal to the vertical distance from the surface of the water to the opening (at any instant v for a falling body is $v = \sqrt{2gh}$). Hence, if the original depth of the water is a, the velocity of the water is $\sqrt{2g(a-x)}$. Therefore, if f is the area of the cross section of the opening, and F, of the surface of the water, the rate at which the surface falls is given by $dx/dt = f\sqrt{2g(a-x)}/F$. (a) Calculate the time required to empty a vertical cylindrical tank of height 10 ft. and radius 2 ft. through a hole of radius 1 in. in the bottom. (b) How long does it take to lower the water 1 ft. when full? (c) How long to empty one half of the water? (Take $g = 32$; t is in seconds.)

11. Using the formula of example 10, calculate the time required to empty a hemispherical bowl of radius 4 ft. if the hole in the bottom is of radius 1 in.

12. For accuracy in actual practice the formula of example 10 is modified by a factor μ, called the coefficient of flow, which depends for its value on the shape of the vessel, the shape of the opening, and other factors. The value of μ ranges from .6 to nearly unity. This corrected formula may be written $dt = \dfrac{1}{\mu f\sqrt{2g}}\displaystyle\int \frac{F\,dx}{\sqrt{a-x}}$. (a) Calculate the general expression for the time required to empty a hemispherical bowl of radius R through an opening in the bottom of cross section f. (b) Evaluate this for the case $R = 4$ ft. and the opening is a 2-inch circular hole for which $\mu = .8$. (Take $g = 32$.) (c) Calculate the time required in (b) to empty one half of the water.

13. Torricelli's law is also applied to finding the rate at which water will escape through a vertical floodgate or spillway in the face of a dam. (a) If the floodgate is rectangular, a feet long and b feet high, and the top is h feet below the surface of the water, calculate the flow, Q, per second. The factor μ must be determined by experiment $\left(\textit{Hint. } Q = \mu a\sqrt{2g}\int_h^{h+b} x^{\frac{1}{2}}\,dx\right)$. (b) Evaluate (a) for the case where the gate is 8 ft. long and 5 ft. deep and the upper edge is 4 ft. below the surface.

14. If in example 13 the escape is over an open weir a feet wide and b feet deep, how fast is the water flowing out when the depth of the overflow is h feet? Check your answer by the formula $Q = \frac{2}{3}\mu A\sqrt{2gh}$, where A is the cross section of water in the weir and h the depth of the overflow.

15. Calculate the width of a weir required to allow 1000 cu. ft. per second to pass when the depth of the overflow is 10 ft. (Take $\mu = .6$.)

16. In organic chemistry there are various applications of the calculus. For example, in the breaking up of cane sugar into dextrose and levulose, in a weak solution, the rate of inversion at any moment is proportional to the amount of unconverted sugar. If a is the original amount of sugar, the inversion formula is $dx/dt = k(a - x)$, where k is a constant, and x is the amount already inverted at the time t. Assume t in minutes, as the process is slow. (a) Calculate the time to convert an amount $y < a$. (b) How long would it require theoretically to transform the entire amount? (c) What part would be transformed in 5 hr. if $k = .0027$ (*Hint.* Solve the result to (a) for y in terms of t). (d) Find the time, if $k = .0027$, required to invert one fourth of the cane sugar; to invert the second fourth; the third fourth.

17. In *bimolecular reaction*, where two molecules M_1 and M_2 of different compounds break up in the presence of some solvent and combine again to form two new molecules M_3 and M_4, differing from each other and from the original molecules, the rate of the reaction is given by (1) $\frac{dx}{dt} = \frac{k(a-x)(b-x)}{v}$, where v, the quantity of the solvent, may be regarded as constant, k is a constant, a and b are the original number of molecules of kinds M_1 and M_2 respectively, or the amounts of M_1 and M_2 we shall say. We can therefore say (2) $dx/dt = k_1(a-x)(b-x)$, where $k_1 = k/v$ and is a constant depending on the concentration, and x is the amount of *each* substance transformed at the time t. (a) Find the time required to transform an amount y of each substance. (b) Find the amount transformed in the time t_1. (c) Suppose in (2) $a = b$, that is, we start with the same number of molecules of each substance, answer (a). (d) Under the conditions of (c) calculate the time required to transform one half of the original substances.

18. Lorenz's law for the rate of cooling of a body is $dT/dt = -aT(1 + bT^{\frac{1}{4}})$, where a and b are constants depending on the body and the surrounding medium, T is the temperature of the body in degrees centigrade, and t is the time in seconds. (a) Calculate the general expression for the time required for a body to cool from a temperature T_1 to T_2. (b) Calculate the time required for a body to cool from 16° to 8° C., using $a = .001$ and $b = .6$.

NOTE. These values of a and b are purely arbitrary.

19. A particle starts at a distance h from the center of the earth and falls toward the earth. The formula for its acceleration after it has fallen through a distance s is $f = gR^2/(h - s)^2$, where g is the gravitational constant and R is the radius of the earth. (a) Calculate its velocity after it has fallen through a distance s from rest; with initial velocity v_0. (b) With what velocity will it reach the earth, starting from rest at a distance $2R$? (c) Answer (b) if it starts with initial velocity v_0. (d) Answer (c) if it starts at a distance h; at an infinite distance.

NOTE. Since $f = dv/dt$ and $v = ds/dt$, we may write $f = v\,dv/ds$.

20. Using the theory and results of example 19, calculate the time required for the particle to fall a distance s, (a) starting from rest; (b) starting with initial velocity v_0.

NOTE. In the result of example 19, (a) put $v = ds/dt$. Integrate by tables or by substitution.

21. If the resistance of the air is assumed to be proportional to the square of the velocity, the law governing the motion of a particle of mass m is $m\,dv/dt = mg - c^2v^2$. Here c is a constant depending on many factors which we need not discuss. (a) Assuming unit mass and $c^2 = .36$, calculate the time required to acquire a velocity of 9 ft. from rest (*Hint.* The limits on v are 0 and 9). (b) Calculate v at the end of 1 sec. (c) Derive the general expression for the time required to acquire a velocity v, starting from rest. (d) Derive the general expression for the velocity acquired in time t. (e) Calculate the distance, s, covered in acquiring a velocity v (*Hint.* Since $dv/dt = ds/dt \cdot dv/ds = v\,dv/ds$, set $v\,dv/ds = g - c^2v^2$ and integrate between the limits 0 and v). (f) What is the maximum speed attainable from rest?

22. (a) Assuming, as in example 21, that the air resistance is proportional to the square of the velocity with which the particle is moving, calculate the time required for a particle projected vertically upwards with an initial velocity v_0, to reach its highest point (*Hint.* In this case $dv/dt = -(g + c^2v^2)$. (b) If $v_0 = 100$ ft./sec. and $c^2 = .0036$, calculate the time required for the particle to come to rest, and compare your answer with v_0/g, the time required when there is no air resistance.

23. If, as in example 22, a particle is projected upwards with an initial velocity v_0, show that the time which elapses until the particle has the velocity v is given by the equation $t = \dfrac{1}{c\sqrt{g}} \arctan \dfrac{c\sqrt{g}\,(v_0 - v)}{g + c^2vv_0}$.

Show also that the velocity at any time t is $v = \dfrac{c\sqrt{g}\,v_0 - g \tan c\sqrt{g}\,t}{c\sqrt{g} + c^2v_0 \tan c\sqrt{g}\,t}$.

Integrate this last expression between the limits $t = 0$ and $t = t$, to find s, the distance covered in t seconds, and show that it can be written in the form

$$s = \frac{1}{c^2} \log \frac{\sqrt{g} \cos c\sqrt{g}\,t + cv_0 \sin c\sqrt{g}\,t}{\sqrt{g}}.$$

Show finally that $s = 1/c^2 \cdot \log \sqrt{(g + c^2 v_0^2)/g}$, where s is the total distance it ascends.

Hint. In integrating $v\,(= ds/dt)$ to obtain s, replace tan by sin/cos and simplify before integrating.

24. (a) Using the methods of example 23, find the height to which a particle will ascend if $v_0 = 100$ ft./sec. and $c^2 = .0036$. (b) Using now the results of example 21, (c) and (e), that is, $t = \frac{1}{2c\sqrt{g}} \log \frac{\sqrt{g} + cv}{\sqrt{g} - cv}$ and $s = \frac{1}{2c^2} \log \frac{g}{g - c^2 v^2}$, calculate the velocity with which the particle returns to the earth and the time required in descent. Compare these results with v_0 and the time of ascent found in example 22, (b).

25. Assume the law that the rate at which water runs out of a hole in the bottom of a tank is proportional to the square root of the height of the surface of the water above the hole. If the tank is of uniform horizontal cross section (as a cylinder, etc.), show that if half the water runs out in 30 min., it will all run out in about 1 hr. 42 min.

26. A vertical cylindrical tank of height h feet and cross section F is filled with water, which is escaping through a hole of area f in the bottom. If a pipe is emptying Q_1 cubic feet per second into the tank, how long will it require for the surface of the water to drop k feet? (See example 10.)

CHAPTER XIV

INTEGRATION A SUMMATION PROCESS. GEOMETRICAL APPLICATIONS

EXERCISE XCV

Plane Area. Rectangular Coördinates. In each of the following examples the student should sketch the curve and rather than substitute in the formula he should draw one of the characteristic rectangles and write down its area. Adding up these rectangles and calculating the limit of the sum as the number becomes infinite, he obtains the required area. This adding is accomplished by integration. The characteristic rectangle may be selected in a vertical or in a horizontal position. That position which leads to the simplest integral should be chosen. The notion of the "summation of elements" is always preferable to the use of formulas, owing to its wide application in geometrical and mechanical problems. Draw a rough figure with a typical element for each problem. (See the adjoining figure.)

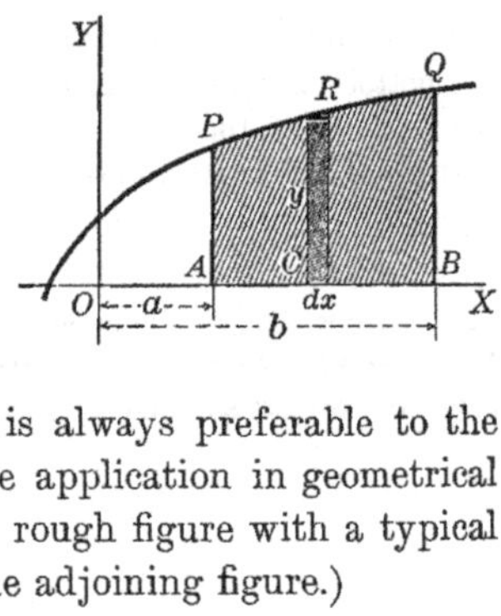

1. For the following curves calculate the area in the first quadrant lying under the arc which extends from the y-axis to the first intercept on the x-axis.

(a) $y = 4 - x^2$.	(h) $y = x^3 - 8x^2 + 15x$.	(o) $y^2 = x^2 - x^3$.
(b) $y = 4 + 3x - x^2$.	(i) $y = (3 - 2x)^2 x$.	(p) $y^2 = x(1 - x)^2$.
(c) $y = 5x - x^2$.	(j) $y = x^4 - 20x^2 + 64$.	(q) $y = e^{x/3} \sin 2x$.
(d) $y^2 = 9 - x$.	(k) $y = x^2(4 - x^2)$.	(r) $y = e^{x/2} \cos 3x$.
(e) $y^2 + y = 2 - x$.	(l) $x = (2 - y)(y + 1)^2$.	(s) $y = 4x^2 - x^3$.
(f) $y^2 - 2y = 3 - x$.	(m) $y^2 = (1 - x)^3$.	(t) $x(y + 1)^2 = 4$.
(g) $y = 9x - x^3$.	(n) $y^2 = x^2 - x^4$.	(u) $y = e^{-x/4} \cos 2x$.

2. Calculate the area bounded by the x-axis, each of the following trigonometric curves and two ordinates separated by a distance equal to a period of the curve. The student should observe that areas below the x-axis obtained by integration are negative, and hence if we calculate the area under one period of the curve, using the entire period for the interval of integration, the result will be the algebraic sum of the areas above and below the axis. To obtain the actual area we must decompose the interval. Thus, if in the interval AB the curve crosses the axis at x_1, we must evaluate the two integrals $\int^{x_1} y\,dx$ and $\int_{x_1}^{b} y\,dx$, then add them *numerically*, independent of sign. The symmetry of the curve should also be referred to in selecting the limits.

(a) $y = \sin 2x$.	(e) $y = \sin^2 x$.	(i) $y = \sin^2 x - \cos x$.
(b) $y = \cos x/2$.	(f) $y = \cos^2 3x$.	(j) $y = \cos 2x + 2\cos x$.
(c) $y = \cos \pi x$.	(g) $y = \sin^2 x + \sin x$.	(k) $y = 3\cos x - 4\sin x$.
(d) $y = \sin 2\pi x/3$.	(h) $y = 2\sin x + \sin 2x$.	(l) $y = 4\cos x + 3\sin x$.

3. A square is formed by the coördinate axes and the point (1, 1). Calculate the ratio of the larger to the smaller of the two areas into which it is divided by each of the following curves:

(a) $y = x^2$.	(e) $x^2 + y^2 = 1$.	(i) $yx + y + x - 1 = 0$.
(b) $y = x^3$.	(f) $x^3 + y^3 = 1$.	(j) $y = e^{-x}$.
(c) $y = x^4$.	(g) $x^{\frac{1}{2}} + y^{\frac{1}{2}} = 1$.	(k) $y = \sin \pi x/2$.
(d) $y^2 = x^3$.	(h) $x^{\frac{2}{3}} + y^{\frac{2}{3}} = 1$.	(l) $y = \tan \pi x/4$.

4. Calculate the area which the given line cuts from the given parabola by subtracting two areas.

(a) $y = x$; $y = 4x - x^2$.	(d) $2y = x + 1$; $y^2 = x + 1$.
(b) $y = x + 7$; $y = 9 - x^2$.	(e) $4y = 4 - x$; $y^2 = 4 - x$.
(c) $y = x + 2$; $y = 4 - x^2$.	(f) $x + y = 1$; $x^{\frac{1}{2}} + y^{\frac{1}{2}} = 1$.

EXERCISE XCVI

Length of a Curve. Rectangular Coördinates. By integrating the expressions for the differential of arc in rectangular coördinates we obtain $s = \int_a^b \sqrt{1 + (dy/dx)^2}\,dx$, or it is sometimes simpler to integrate the second form, $s = \int_c^d \sqrt{1 + (dx/dy)^2}\,dy$.

1. In the following curves calculate the length of the arc joining the two given points.

(a) $y^2 = x^3$; (0, 0), $\left(\frac{5}{9}, 5\sqrt{5}/27\right)$.	(c) $y^2 = 4x - x^2$; (0, 0), $\left(1, \sqrt{3}\right)$.
(b) $y = \frac{2}{3}(4 - x)^{\frac{3}{2}}$; $(0, \frac{16}{3})$, (4, 0).	(d) $3y^2 = 8x^3$; (0, 0), $\left(1, 2\sqrt{2/3}\right)$.

(e) $y^2 = 20x$; $(0, 0), (4, 4\sqrt{5})$.
(f) $6y = x^2$; $(0, 0)$, $(4, 8/3)$.
(g) $3y = x^3$; $(0, 0)$, $(1, 3)$.
(h) $y = \sqrt{8}\log x$; $(1, 0)$, $(\sqrt{8}, \sqrt{2}\log 8)$.
(i) $y = \log\sec x$; $(0, 0)$, $(\pi/3, \log 2)$.

2. (a) $3x^2 = y^3$; $(1/\sqrt{3}, 1)$, $(40\sqrt{15}/3, 20)$.
(b) $4x = y^2 - 4$; $(0, -2)$, $(0, 2)$.
(c) $x = (e^{y/2} + e^{-y/2})$ from $y = 0$ to $y = 2$.

EXERCISE XCVII

Volume of a Solid of Revolution. Consider the section of a solid of revolution made by a plane perpendicular to its axis. Denote the area of this section by A. If we multiply this by da, a differential distance *along the axis perpendicular to the plane,* we obtain the volume $A\,da$, of a characteristic cylindrical element. Using the notion of summation gives the integral

$$\int A\,da,$$

which is the required volume when evaluated for the proper limits. This method of "setting up the element" should be applied to each of the following examples. Draw a rough figure showing the cylindrical element in each case.

The student should observe that A in general has the form πr^2, where r is the distance to the axis from a point in the generating curve. In case the area of the section is ring-shaped (bounded by two concentric circles) A takes the form $\pi(r_1^2 - r_2^2)$. When the x-axis is the axis of revolution, the radius reduces to the ordinate of a point on the curve, giving the more familiar formula $\pi\int_a^b y^2\,dx$, or using the y-axis as the axis of revolution, we obtain $\pi\int_c^d x^2\,dy$.

1. Calculate the volumes obtained by revolving the areas indicated about the x-axis. The boundary curves and lines are given.

(a) $y = x^3$, $y = 0$, $x = 1$.
(b) $y^2 = x^3$, $y = 0$, $x = 5$.
(c) $y = \sin x$, $y = 0$.
(d) $y = \sin 2x$, $y = 0$.
(e) $y = x^2 - 4x$, $y = 0$.
(f) $y = e^{ax}$, $y = 0$, $x = 0$, $x = b$.

2. In these examples revolve the areas about the y-axis and compute the volumes.

(a) $x^2 = 4y - 2y^2$, $x = 0$.
(b) $4y - y^2 + 4x = 0$, $x = 0$.
(c) $x^2 - y^2 + 4 = 0$, $x = 0$, $x = 2$.
(d) $x^2 = y^3 - 4y$.

3. In these examples the area between the parametric curve and the x-axis is to be revolved about the x-axis.

(a) $x = 2 - t$, $y = t^2 - 4$.
(b) $x = t + 1$, $y = t^2 - 4t$.
(c) $x = t^2$, $y = 4t - t^3$.
(d) $x = t^3/3$, $y = 3t - t^2$.

4. Calculate the volume of the solid generated by revolving about each of the following lines the area which they cut from the corresponding curves:

(a) $y = 3$; $y = 3 + 2x - x^2$.
(b) $x = 3$; $x = 4y - y^2$.
(c) $x = 3$; $y^2 - 2y + x - 3 = 0$.
(d) $yx = 4$; $y^2 = x^3$.
(e) $x = 5$; $x^2 - y^2 = 16$.
(f) $y = 1$; $x^2 + y^2 = 4$.
(g) $y = x$; $y = x^2$.
(h) $y = x$; $y = 4x - x^2$.
(i) $y = x + 7$; $y = 9 - x^2$.
(j) $2y - x - 1 = 0$; $y^2 = x + 1$.
(k) $x + y = 1$; $x^{\frac{1}{2}} + y^{\frac{1}{2}} = 1$.
(l) $x + y = 5$; $xy = 4$.

5. By revolving it about the axis indicated, find the volume generated by the finite area bounded by the following pairs of curves:

(a) y-axis, $xy = 4$, $x + y = 5$.
(b) y-axis, $y^2 = x$, $x + y = 2$.
(c) x-axis, $y = 2 + x - x^2$, $y = 2 - x$.
(d) y-axis, $x^2 - 4x + y = 0$, $y = 0$.
(e) y-axis, $y^2 = x^3$, $y^2 = 2 - x$.
(f) y-axis, $y^3 = 27x$, $3x = 4y - y^2$.
(g) x-axis, $y = (4 + x^2)/x$. $y = 5$.
(h) x-axis, $y = x^2$, $y = 4x - x^2$.
(i) x-axis, $y = x^3$, $y^2 = x$.
(j) y-axis, $y = x^3$, $y^2 = x$.
(k) x-axis, $y = x^2$, $y^2 = x$.
(l) x-axis, $y = \sin \pi x/2$, $y = x^2$.

6. (a) The area bounded by the curve $y = \sin x$ and the lines $x = \pi/2$ and $y = 0$ is revolved about the x-axis. Find the volume.

(b) The area bounded by the same arc of $y = \sin x$, the y-axis, and the line $y = 1$ is revolved about the y-axis. Find the volume. (b) is a good illustration of a case where it is best to replace the differential, that is, to change the independent variable to x.

7. Given the curve $x = t^2$, $y = 4t - t^3$. Find (a) the area of the loop and (b) the volume generated by the area inside the loop when revolved about the x-axis.

8. Find the volume of the solid generated by revolving about the x-axis the area bounded by $y = e^{ax}$ and the coördinate axes.

9. Revolve the area between the two parabolas $y^2 = 4x$ and $y^2 = 5 - x$ about each axis and calculate the respective volumes.

10. Revolve about the polar axis the part of the cardioid $\rho = 4 + 4\cos\theta$ between the lines $\theta = 0$ and $\theta = \pi/2$ and compute the volume.

EXERCISE XCVIII

Area of a Surface of Revolution. Regarding the axis of x as the axis of revolution, the characteristic element becomes the lateral surface of a frustum of a cone whose slant height is the differential of arc and whose mean radius is the ordinate to the curve; hence the element of surface $= 2\pi y\,ds$. Using the y-axis as the axis of revolution, the formula becomes $2\pi x\,ds$, or, generally, $2\pi r\,ds$. In any case, ds may be replaced by either of the forms $\sqrt{1 + (dy/dx)^2}\,dx$, or $\sqrt{1 + (dx/dy)^2}\,dy$, according as the one or the other leads to the simpler form to integrate. *The fact that the y-axis is the axis of revolution does not demand the second form, as many students suppose.*

1. In each of the following curves calculate the area of the surface generated by revolving about the x-axis the arc connecting the given points:

(a) $9y = x^3$; $(0, 0)$, $(2, \frac{8}{9})$.
(b) $y^2 = 4x$; $(0, 0)$, $(3, 2\sqrt{2})$.
(c) $y^2 = 9x$; $(0, 0)$, $(4, 6)$.
(d) $y^2 = 24 - 4x$; $(3, 2\sqrt{2})$, $(6, 0)$.
(e) $x^2 + y^2 = 4$; $(1, \sqrt{3})$, $(2, 0)$.
(f) $6y = x^2$; $(0, 0)$, $(4, \frac{8}{3})$.

2. Calculate by integration the surface generated by revolving about the x-axis one arch of the curve $y = \sin x$.

3. Calculate the entire surface of a solid generated by revolving about the x-axis the area bounded by the two parabolas $y^2 = 4x$; $y^2 = 3 + x$.

EXERCISE XCIX

Plane Area and Length of Arc. Polar Coördinates. The student should plot the polar curve for each example to assist him in determining the proper limits for the variable angle θ. A characteristic element of area, the area between two radii vectores with an angle $d\theta$ between them, — hence a circular sector of area

$\frac{1}{2}\rho^2 d\theta$, — should be marked on each figure, and the notion of summation should be applied as before. The element of arc in polar coördinates is $\sqrt{\rho^2 + (d\rho/d\theta)^2}\,d\theta$, therefore

$$s = \int_{\theta_1}^{\theta_2} \sqrt{\rho^2 + (d\rho/d\theta)^2}\,d\theta.$$

1. Calculate the total area inclosed by each of the following curves:

(a) $\rho^2 = 4 \sin 2\theta$. (e) $\rho = 1 + \cos\theta$. (i) $\rho = \sin^2 \theta/2$.
(b) $\rho^2 = 9 \cos 2\theta$. (f) $\rho^2 = \sin\theta + 1$. (j) $\rho = \cos^2 \theta/2$.
(c) $\rho = \cos 2\theta$. (g) $\rho = 3 - \sin\theta$. (k) $\rho = \frac{1}{2} + \cos 2\theta$.
(d) $\rho = \sin 3\theta$. (h) $\rho = 2 - \cos\theta$. (l) $\rho = \sin 2\theta - \frac{1}{2}$.
(m) $\rho = \sin 3\theta + 2$. (o) $\rho = \cos 3\theta - \cos\theta$.
(n) $\rho = 2 + \cos 3\theta$. (p) $\rho = \cos 3\theta - 2\cos\theta$.

2. Calculate the ratio of the area of the larger to the area of the smaller loop of the following curves:

(a) $\rho = \frac{1}{2} + \cos\theta$. (c) $\rho = \frac{1}{2} + \cos 2\theta$
(b) $\rho = \sin^3 \theta/3$. (d) $\rho = \frac{1}{2} + \sin 3\theta$.

3. Calculate the area which the curves in each of the following pairs have in common. Each example will involve the evaluation of two definite integrals having one limit in common determined by the points of intersection of the two curves. The sum of these two integrals will give the required answer.

(a) $\rho = \cos\theta$, $\rho = \sin\theta$.
(b) $\rho^2 = \cos 2\theta$, $\rho^2 = \sin 2\theta$.
(c) $\rho = 1$, $\rho = 1 + \cos\theta$.
(d) $\rho = \sin\theta$, $\rho = 1 - \cos\theta$.
(e) $\rho = 3\cos\theta$, $\rho = 1 + \cos\theta$.
(f) $\rho = \sqrt{3}\sin\theta$, $\rho = 1 + \cos\theta$.
(g) $\rho = 1$, $\rho^2 = 2\cos 2\theta$.
(h) $\rho = \sqrt{6}\cos\theta$, $\rho^2 = 9\cos 2\theta$.
(i) $\rho = \sqrt{2}\sin\theta$, $\rho^2 = \cos 2\theta$.
(j) $\rho = \sqrt{2}\cos\theta$, $\rho^2 = \sqrt{3}\sin 2\theta$.
(k) $\rho = (\cos\theta)/\sqrt{3}$, $\rho = \cos 2\theta$.
(l) $\rho = \sqrt{2/3}\sin 2\theta$, $\rho^2 = \cos 2\theta$.

4. Find the total length of each of the following polar curves:

(a) $\rho = \cos^2 \theta/2$. (b) $\rho = \sin^3 \theta/3$. (c) $\rho = \sin^4 \theta/4$. (d) $\rho = 1 + \cos\theta$.

5. Given the parabola $\rho = 2/(1 + \cos\theta)$. (a) Find the length of arc to the right of the line $\theta = \pi/2$. (b) Find the area cut from the parabola by a line through the pole perpendicular to the polar axis.

6. (a) Find the length of arc of the entire cardioid $\rho = 2(1 - \cos\theta)$. (b) Rotate the area bounded by this cardioid about the polar axis and calculate the surface generated.

EXERCISE C.

Volumes of Miscellaneous Solids. Just as we obtained the circular sections of the solids of revolution in a preceding exercise, so we may obtain various sections — elliptical, triangular, rectangular, etc. — of miscellaneous solids, and multiplying the area of such a section by the differential of that variable, say x, to whose axis the section is perpendicular, we obtain the volume of a characteristic element. It is essential that the area of this element be a function of x. Applying the notion of summation we arrive at the volume integral for any particular solid. For example, consider the volume cut from the paraboloid $4y^2 + 9z^2 = 4x$ by the plane $x = 9$. Any element perpendicular to the x-axis is an elliptic cylinder of semiaxes $\sqrt{x}$ and $\frac{2}{3}\sqrt{x}$, respectively (and therefore of area $\frac{2}{3}\pi x$), and of thickness dx.

Hence the volume $= \int_0^9 \frac{2}{3}\pi x\,dx = 27\pi.$

1. Calculate the volumes bounded by the following quadric surfaces and the given planes:

(a) $z = x^2 + 4y^2$; $z = 1$.
(b) $4x^2 + 9z^2 + y = 0$; $y + 1 = 0$.
(c) $x^2 + 4y^2 = 1 + z^2$; $z \pm 1 = 0$.
(d) $25y^2 + 9z^2 = 1 + x^2$; $x - 2 = 0$.
(e) $x^2 + 4y^2 + 9z^2 = 1$.
(f) $z^2 = x^2 + 9y^2$; $z \pm 1 = 0$.

2. The double ordinate of the ellipse $x^2 + 4y^2 = 4$ serves respectively as

(a) the side of a square,
(b) the diagonal of a square,
(c) the base of an equilateral triangle,
(d) the altitude of an equilateral triangle,
(e) the diagonal of a regular hexagon.

Calculate the volume of the solid generated by the area of each of these figures as it moves, the plane always remaining perpendicular to the x-axis.

3. Given the parabola $z = 4 - x^2$ in the XZ-plane and the circle $x^2 + y^2 = 4$ in the XY-plane. From each point on the parabola lying above the circle two lines are drawn parallel to the YZ-plane to meet the circle. Calculate the volume of the wedge-shaped solid thus formed.

4. The double ordinates of the circle $x^2 + y^2 = 25$ serve as the bases of isosceles triangles with a constant altitude of 10. Calculate the volume of the solid made up of all such triangles standing at right angles to the plane of the circle.

5. The plane sections of a solid by planes perpendicular to the x-axis are ellipses whose major and minor axes are the double ordinates of the curves $y^2 = x$ and $z^2 = x^3$ respectively. Calculate the volume cut from the solid by the plane $x = 2$.

6. A football is 16 in. long and a plane section containing a seam is an ellipse the shorter diameter of which is 8 in. Find the volume, (a) if the leather is so stiff that every cross section is a square ; (b) if the cross section is a circle.

7. An isosceles right triangle moves with its plane perpendicular to the XY-plane and with its hypotenuse always parallel to the y-axis. One end of the hypotenuse moves on the line $y = 3x$ and the other end on the curve $x^2 + y = 0$. Find the volume generated when the triangle moves from $x = 0$ to $x = 2$.

8. The cap of a square pillar is of the form of the volume common to two equal semicircular cylinders of radius a, the axes of the cylinders intersecting at right angles. Find the volume of the cap.

Hint. The horizontal cross sections are all squares.

CHAPTER XV

APPROXIMATE INTEGRATION — VARIOUS METHODS

EXERCISE CI

The Definite Integral. Approximate Evaluation by the Use of Series. Integrate the following functions by the method of Exercise LXXXIX. The interval of integration is in each case so selected that the resultant series will converge rapidly for both limits. The student should use a sufficient number of terms to have his answer correct to at least two decimal places.

1. $\int_0^1 \sqrt{1+x^3}\,dx.$
2. $\int_0^1 \sqrt[3]{1+x^2}\,dx.$
3. $\int_0^1 \sqrt[3]{1+x^3}\,dx.$
4. $\int_0^1 \sqrt{1+x^4}\,dx.$
5. $\int_{.1}^{.2} \frac{dx}{\sqrt{1-x^3}}.$
6. $\int_{.1}^{.2} \frac{dx}{\sqrt{1+x^4}}.$
7. $\int_0^1 \frac{\sin x\,dx}{x}.$
8. $\int_0^1 \cos x^2\,dx.$
9. $\int_0^{\pi/6} \sqrt{\cos x}\,dx.$
10. $\int_0^{\pi/6} \sqrt{\sec x}\,dx.$
11. $\int_0^{\pi/6} \sqrt{\sin x}\,dx.$
12. $\int_0^1 \sqrt{\text{arc tan}\,x}\,dx.$

EXERCISE CII

The Definite Integral. Simpson's Rule. It is evident that the method of integration by series employed in Exercise CI cannot be applied in evaluating a definite integral unless the resultant series converges for the values of the limits. This, in general, will not be so, and even though the series be convergent for these values, it may be necessary to compute too many terms before arriving at a sufficiently accurate result.

The student will recall, however, that the definite integral in general can be represented by the area under the graph of the function to be integrated. Any method, therefore, for the approximate calculation of this area will serve equally well in evaluating the definite integral *whatever it may represent.* The area under a curve can be computed with considerable accuracy by dividing it into small parts by equally spaced ordinates and then applying either the trapezoidal or Simpson's rule, which follow respectively:

$$A_T = \Delta x(y_1/2 + y_2 + y_3 + \cdots + y_{n-1} + y_n/2).$$

$$A_S = \Delta x/3 \cdot (y_1 + 4\,y_2 + 2\,y_3 + 4\,y_4 + 2\,y_5 + \cdots + 4\,y_{n-1} + y_n).$$

In both rules, Δx represents the distance between two successive ordinates and the y's are the ordinates. No restriction is placed on n in the first rule, but to apply Simpson's rule the number of ordinates must be odd. The larger the number of ordinates between the limits, the greater the accuracy. For the same number of ordinates Simpson's rule gives the more accurate results in most cases. The following integrals are intended for evaluation by these methods, the number of ordinates to be used being indicated in brackets in the earlier examples.

1. $\int_0^2 \sqrt{8 + x^3}\,dx$; (3).

2. $\int_0^2 \sqrt[3]{7 + 5x^2}\,dx$; (3).

3. $\int_0^1 \sqrt{1 + x^4}\,dx$; (4).

4. $\int_0^1 \sqrt[3]{1 + x^2}\,dx$; (4).

5. $\int_4^{10} 1/\log_{10} x\,dx$; (7).

6. $\int_1^{11} \sqrt{\log_{10} x}\,dx$; (11).

7. $\int_1^6 \sqrt{1 + \log_{10} x}\,dx$; (11).

8. $\int_1^6 1/(1 + \log_{10} x)\,dx$; (11).

9. $\int_0^{\pi/2} \sqrt{\cos\theta\,d\theta}$; (7).

10. $\int_0^{\pi/2} 1/\sqrt{4 + \sin\theta}\,d\theta$; (7).

11. $\int_0^{\pi/4} \sqrt{1 + \tan x}\,dx$; (5).

12. $\int_1^2 \cos(\log x)\,dx$; (5).

13. $\int_0^{\pi/2} \sqrt{1 + \sin^2\theta}\,d\theta$; (7).

14. $\int_0^{\pi/4} \sqrt{1 + 3\sin^2 2\,\theta}\,d\theta$; (7).

15. $\int_0^{\pi/2} \frac{dx}{1 + \cos x} dx$; (3).

16. $\int_1^2 \frac{dx}{1 + \log x}$; (10).

17. $\int_1^2 \sqrt{1 + \log x}\,dx$; (5).

18. $\int_0^{\pi/2} \sqrt{\sin x}\,dx$; (6).

19. $\int_0^1 \sin e^x\,dx$; (5).

20. $\int_1^2 \tan(\log x)\,dx$; (5).

21. Find, correct to two decimal places, the length of one arch of the curve $y = \sin x$.

22. Calculate approximately the total perimeter of the four-leaved rose $\rho = \sin 2\theta$.

23. Calculate the approximate length of arc of the curve $4y = x^3$ from the origin to (2, 2).

24. Given the curve $y = e^{-x/4}$. (a) Calculate the approximate length of arc from $x = 0$ to $x = 4$. (b) Revolve this arc about the x-axis and calculate the approximate area of surface generated.

25. Revolve the curve $y^2 = x^3 + 1$ from $x = 0$ to $x = 4$ about the x-axis, and calculate the surface generated.

26. The curve $y = \log_{10} x$ from $x = 1$ to $x = 5$ is revolved about the x-axis. Calculate the surface generated.

27. The following example illustrates the accuracy of Simpson's rule. Integrate $\int_0^1 1/(1 + x^2)\,dx$ exactly. Then calculate the value of the integral by Simpson's rule, using ten intervals and carrying the result to seven or more decimal places. Equate the two values and solve for π. See by comparing with the value of π given in your textbook that it is correct to seven decimal places.

EXERCISE CIII

Geometrical Integrals. Review. It is intended that in each of the following assignments the student shall find the area inclosed by the given curve or curves and shall calculate the *total* perimeter (both curvilinear and rectilinear) of the area, the volume of the solid obtained by revolving the area about the x-axis, and the area of the *total* surface (both curved and plane) of that solid. When necessary, reference should be made to the table of integrals, and other numerical tables should be used freely in simplifying the results. In most cases it will be necessary to use Simpson's rule (Exercise CII) in the evaluation of one or more of the integrals arising in each of the examples.

1. $6y = x^2$, $x = 4$, $y = 0$.	**2.** $y = +\sqrt{20x}$, $x = 4$, $y = 0$.	**3.** $y = +\sqrt{24 - 4x}$, $x = 3,\ x = 6$, $y = 0$.

4. $y = +\sqrt{4 - x^2}$,
$x = 0$,
$x = 1$,
$y = 0$.

5. $y = x^3$,
$x = 2$,
$y = 0$.

6. $y = +\sqrt{x^3}$,
$x = 5$,
$y = 0$.

7. $y = \log x$,
$x = 2$,
$y = 0$.

8. $y = \sin x$,
$x = \pi/2$,
$y = 0$.

9. $y = e^x$.
$x = 0$,
$x = 1$,
$y = 0$.

10. $y = \frac{1}{2}(e^x + e^{-x})$,
$x = 0$,
$x = 1$,
$y = 0$.

11. $x^{\frac{1}{2}} + y^{\frac{1}{2}} = 1$,
$x = 0$,
$y = 0$.

12. $x^{\frac{2}{3}} + y^{\frac{2}{3}} = 1$.

13. $\rho = 1 + \cos\theta$.

14. The area bounded by the x-axis and one arch of the cycloid, $x = t - \sin t$, $y = 1 - \cos t$. Find also the volume if this area is revolved about a tangent at the highest point.

CHAPTER XVI

MULTIPLE INTEGRATION

EXERCISE CIV

Definite Double and Triple Integrals. The student should not forget that when integrating with respect to any one variable, all other variables must be regarded as constants. The order of integration and the association of integral signs with differentials used in this book is indicated in the following scheme:

$$\int_e^f \boxed{\int_c^d \boxed{\int_a^b f(x,\,y,\,z)\,dx}\,dy}\,dz,$$

the operation in the inner rectangle being the first to be performed.

1. $\displaystyle\int_0^2 \int_{\sqrt{2x}}^{x - x^2} dy\,dx.$

2. $\displaystyle\int_0^{\sqrt{2}} \int_{x^3 - x}^{3x - x^3} dy\,dx.$

3. $\displaystyle\int_0^2 \int_{-\frac{x^2}{4}}^{\frac{3x - 2x^2}{2}} dy\,dx.$

4. $\displaystyle\int_{-2}^1 \int_{y^2 + 2y}^{4 - y^2} dx\,dy.$

5. $\displaystyle\int_{-1}^2 \int_{2x^2 - 2}^{x^2 + x} x\,dy\,dx.$

6. $\displaystyle\int_{-1}^2 \int_{2x^2 - 2}^{x^2 + x} y\,dy\,dx.$

7. $\displaystyle\int_{-1}^2 \int_{x^3 + x}^{\sqrt{12x + 12}} y\,dy\,dx.$

8. $\displaystyle\int_{-2}^{+2} \int_{\sqrt[3]{4y}}^{y - y^2 + 4} y\,dx\,dy.$

9. $\displaystyle\int_0^1 \int_{\sqrt{y}}^{2 - y} y^2\,dx\,dy.$

10. $\displaystyle\int_0^1 \int_{x^2}^{2 - x} x^2\,dy\,dx.$

11. $\displaystyle\int_0^1 \int_{x^2}^{2 - x} y^2\,dy\,dx.$

12. $\displaystyle\int_0^1 \int_{\sqrt{y}}^{2 - y} x^2\,dx\,dy.$

13. $\displaystyle\int_0^1 \int_0^{\sqrt{2x - x^2}} \int_0^{2 - x} dz\,dy\,dx.$

14. $\displaystyle\int_0^1 \int_{y^2}^{2 - y} \int_0^{2 - y - x} dz\,dx\,dy.$

15. $\displaystyle\int_0^1 \int_{y^2}^1 \int_0^{1 - x} x\,dz\,dx\,dy.$

16. $\displaystyle\int_0^1 \int_0^{1 - x} \int_0^{1 - y^2} z\,dz\,dy\,dx.$

EXERCISE CV

Plane Area by Double Integration. Rectangular Coördinates. The student should always draw an accompanying figure for each example to help him ascertain the proper limits for integration. He should also represent one of the characteristic rectangles or *checks* ($dx\,dy$) within the required area and should sum up checks parallel to a coördinate axis whenever possible in such a manner that the resultant *strip*, independent of the position of the first check, will extend from one curve to the other (that is, not meeting the same curve twice). This is important, for in this way only *one* double integral need be evaluated. (See the accompanying figure for the area bounded by $y^2 = x^3$, $y = x$.) These directions should be applied in calculating the finite area bounded by each of the following pairs of curves.

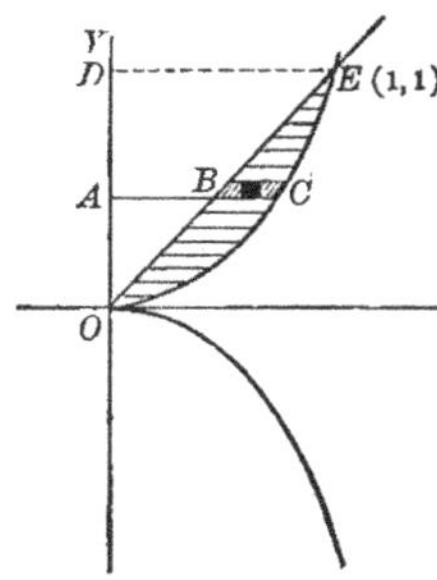

1. $y + x = 1$, $y^2 = x + 1$.
2. $y + x = 1$, $y = x^2 - x$.
3. $y = x$, $y = 4x - x^2$.
4. $y = x + 7$, $y = 9 - x^2$.
5. $y^2 = 4x$, $2x - y - 4 = 0$.
6. $x = y^2 - 2y$, $y = x$.
7. $xy = 4$, $x + y = 5$.
8. $y(2 - x) = 6$, $y + 2x = 8$.
9. $y^2 = 4 - x$, $y^2 = 4 - 4x$.
10. $y^2 = 5 - x$, $y^2 = 4x$.
11. $y = 2x - x^2$, $y = 3x^2 - 6x$.
12. $y^2 = 4 - x$, $y^2 + 2y = x$.
13. $4y + x^2 = 0$, $2y - 3x + 2x^2 = 0$.
14. $y = 2x - x^2 + 1$, $3y = 3 - x^2$.
15. $y = x^2 + x$, $y = 2x^2 - 2$.
16. $y = x^2 - 4x$, $y = 6 - x^2$.
17. $y^2 = 4 + 2x$, $y^2 = 4 - x$.
18. $9y = (x + 3)^2$, $y = (x - 1)^2$.
19. $4y = 3x^2$, $2y^2 = 9x$.
20. $2x^2 = 9y$, $4x = 3y^2$.
21. $y^2 = 2x$, $y = x - x^2$.
22. $y = x^2 + x$, $y^2 = 12(1 + x)$.
23. $y = x^3 - 3x$, $4y = x^3$.
24. $y = x^3 - 2x$, $y = 6x - x^3$.

25. $y = x^3 - x,$
$y = 3x - x^3.$

26. $y = x^3 - 2x,$
$3y = x^3.$

27. $x^2 + y^2 - 10x = 0,$
$y^2 = 24 - 4x.$

28. $x^2 + y^2 = 25,$
$4x^2 - 9y = 0.$

29. $y^2 = 4x,$
$5y = 12 - 2x^2.$

30. $4y = x^3,$
$x = y - y^2 + 4.$

31. Calculate the area bounded by the y-axis and the two curves $y = \sin x$ and $y = \cos x$.

32. Find by double integration the area inside the circle $x^2 + y^2 = 25$, between the line $x = 0$ and the parabola $4y^2 = 9x$.

33. Calculate the ratio of the larger to the smaller of the two finite areas bounded by the following pairs of curves.

(a) $x^2 + y^2 = 4,$
$x = 1.$

(b) $3y^2 + x^2 = 4,$
$3y - x = 2.$

(c) $4y = x^2,$
$4y = x^3 - 2x.$

(d) $y = 5x,$
$y = x^2(x + 4).$

(e) $y = x(8 - x),$
$y = x(4 - x)^2.$

(f) $y = 2\sin x,$
$y = 2\cos x/2.$

(g) $x^2 + y^2 = 12,$
$y^2 = 4x.$

(h) $x^2 + y^2 = 25,$
$x^2 = 4y - 7.$

(i) $y = 4x - x^3,$
$2y = x^2 - 4.$

EXERCISE CVI

Plane Area by Double Integration. Polar Coördinates. The student should review the method for finding the coördinates of the point of intersection of two polar curves to facilitate the calculation of the limits. Owing to the fact that the equations of curves in polar coördinates in general give complicated results when solved explicitly for θ, it follows that it is better to choose constant limits for that variable and perform the first partial integration with respect to the variable ρ. Bearing this in mind, the student should draw the characteristic check, which we treat as a rectangle, of sides $\rho d\theta$ and $d\rho$, and hence of area $\rho d\rho d\theta$, within the required area and then sum up checks to form sectors which, if possible, reach from one curve to the other. If he will refer to the areas found under example 3, Exercise XCIX, he will notice that this is not always possible and that such areas are best calculated by two single integrals.

The following ten curves will be referred to by letter in the examples of this exercise:

Cardioid, K; $\rho = 1 + \cos\theta$.
Parabola, P; $\rho = 1/(1 + \cos\theta)$.
Straight line, L; $\rho = 3/4 \cos\theta$.

Circles

C_1; $\rho = 1/2$. C_2; $\rho = 1$. C_3; $\rho = 3/2$.
C_4; $\rho = 2$. C_5; $\rho = \cos\theta$. C_6; $\rho = \frac{4}{3}\cos\theta$.
C_7; $\rho = 3\cos\theta$.

Find the area of the following:

1. C_4 outside of C_1.
2. C_3 to right of L.
3. C_7 outside of C_3.
4. C_7 outside of C_5.
5. C_6 outside of C_5.
6. C_5 outside of C_1.
7. C_7 to right of L.
8. C_5 to right of L.
9. K to right of L.
10. K outside of C_2.
11. K outside of C_3.
12. K outside of C_1.
13 C_1 outside of K.
14. C_2 outside of K.
15. C_3 outside of K.
16. C_7 outside of K.
17. C_2 outside of P.
18. C_4 outside of P.
19. C_6 outside of P.
20. K outside of P.
21. K between L and P.

22. Calculate that part of the area of the circle $\rho = \cos\theta + \sin\theta$ lying outside the circle $\rho = 1$.

23. Calculate that part of the area of the circle $\rho = \sin\theta$ lying outside the cardioid $\rho = 1 - \cos\theta$.

24. Calculate that part of the area of the lemniscate $\rho^2 = 2a^2\cos 2\theta$ which lies outside the circle $\rho = a$.

25. Calculate that part of the area of the lemniscate $\rho^2 = 9\cos 2\theta$ which lies outside the circle $\rho = \sqrt{6}\cos\theta$.

26. Calculate that part of the area of the four-leaved rose $\rho = a\cos 2\theta$ lying outside the circle $\rho = a/2$.

27. Calculate that part of the area of the cardioid $\rho = 4(1 + \cos\theta)$ which lies outside the parabola $\rho = 3/(1 - \cos\theta)$.

28. Calculate the area of the smaller segment which $\rho\sin\theta = a$ cuts from $\rho = 2a$.

EXERCISE CVII

Volumes by Triple Integration. In this case the element of volume is a rectangular parallelopiped with edges Δx, Δy, and Δz or dx, dy, and dz in the limit. Observe the adjoining figure. The student should draw this element in each figure; then begin by summing up elements parallel to one of the axes so that the resulting *column* will extend without interruption from one bounding surface to another, independent of the position of the element within the volume; then sum up columns to get a *slice*; finally, sum up the slices to get the volume. The outline of a single example, where the student must supply the figure, will be of value if carefully studied. For example, find by triple integration the volume cut by the plane $z = 4$ from the paraboloid $z = x^2 + 2y^2$.

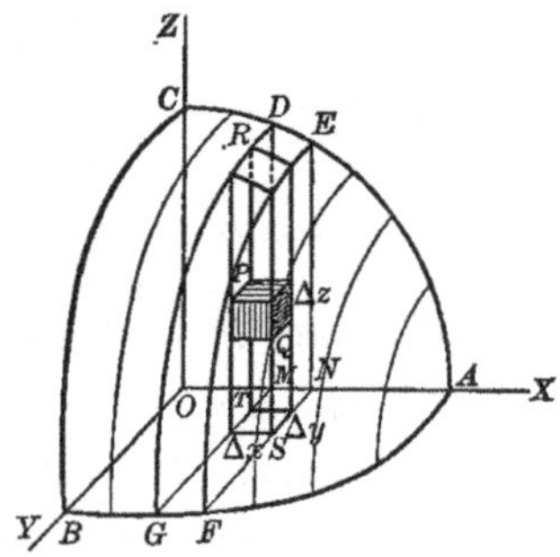

Solution:

$$\frac{1}{4}V = \int_0^2 \int_0^{\sqrt{\frac{4-x^2}{2}}} \int_{x^2+2y^2}^{4} dz\,dy\,dx = \int_0^2 \int_0^{\sqrt{\frac{4-x^2}{2}}} (4 - x^2 - 2y^2)\,dy\,dx$$

$$= \frac{\sqrt{2}}{3} \int_0^2 (4 - x^2)^{3/2} dx = \sqrt{2}\,\pi.$$

Hence $V = 4\sqrt{2}\,\pi$. The limits of x and y were obtained from the *curve of intersection* of the bounding surface and plane.

1. Find by triple integration the volume of the tetrahedron formed by the plane $x/a + y/b + z/c = 1$ and the three coördinate planes.

2. Find the volume bounded by the planes $z = 0$, $y = 0$, $z = x$ and the cylinder $x^2 + y^2 = 9$.

3. Find the volume in the first octant common to the two cylinders $x^2 + y^2 = 4$ and $x^2 + z^2 = 4$.

In the remaining examples the solids are completely bounded by the given surfaces and planes. Find the volume by triple integration.

4. $y^2 = x$, $x + z = 1$, $y = 0$, $z = 0$.

5. $y^2 = x$, $x + y + z = 2$, $y = 0$, $z = 0$.

6. $y^2 + z = 1$, $x + y = 1$, $x = 0$, $z = 0$.

7. $x^2 + z = 1$, $y^2 + z = 1$, $x = 0$, $y = 0$, $z = 0$.

8. $x^2 + y^2 - 2x = 0$, $2x + z - 2 = 0$, $y = 0$, $z = 0$.

9. $y^2 - 2x = 4$, $x + z = 1$, $z = 0$.

10. $x^2 + y^2 = 4$, $z + y = 3$, $z = 0$.

11. In cylindrical coördinates the volume of a solid is given by the triple integral $\iiint \rho d\rho d\theta dz$. Use this to calculate the volume of a cone of radius 2 and altitude 3.

12. Use cylindrical coördinates to calculate the volume of a paraboloid of revolution, the radius of the base and the altitude each being 2 units.

13. In spherical coördinates the volume of a solid is given the triple integral $\iiint \rho^2 \sin\phi \, d\theta d\phi d\rho$. Use this integral to find the volume of a sphere. (The limits for $\frac{1}{8}V$ are $(0, \pi/2)$, $(0, \pi/2)$, and $(0, r)$ respectively.)

14. Use spherical coördinates to find the part cut from a circular cone of 60° vertex angle by a sphere of radius 6 in. The vertex of the cone is at the center of the sphere.

CHAPTER XVII

GENERAL APPLICATIONS

EXERCISE CVIII

Fluid Pressure. This exercise is concerned with the calculation of pressure against flat areas which are immersed at right angles to the surface of the fluid. The problem is inserted as another application of summation, the area in question being divided into strips parallel to the fluid surface. Calculate the pressure on one characteristic strip as a function of its distance below the surface and obtain the total pressure by integration. The pressure on a strip is the area of the strip multiplied by W times the depth below the surface, where W is the weight of a cubic unit of the liquid. That is, if x is the length of a typical strip, then xdy is the area and $Wxydy$ the pressure, y being the depth below the surface. The typical formula is $W\int_a^b xydy$, but the student should set up the element in each example. Fluid pressure may also be calculated as a double integral.

1. Assuming the weight of one cubic unit of fluid as W and regarding the x-axis as the fluid level to which the y-axis is perpendicular, calculate the pressure on the areas formed by joining with straight lines the following sets of points in the order given:

(a) $(0, 0), (3, 0), (3, -2), (0, -2), (0, 0)$.
(b) $(0, 0), (3, 0), (0, -2), (0, 0)$.
(c) $(0, 0), (3, -2), (0, -2), (0, 0)$.
(d) $(-1, 0), (2, 0), (0, -3), (-1, 0)$.
(e) $(0, 0), (3, -2), (-1, -2), (0, 0)$.
(f) $(-2, 0), (3, 0), (3, -2), (0, -2), (-2, 0)$.
(g) $(-1, 0), (4, 0), (2, -2), (0, -2), (-1, 0)$.

(h) (0, 0), (2, − 2), (0, − 3), (0, 0).
(i) (0, 0), (1, − 1), (0, − 2), (− 1, − 1), (0, 0).
(j) (0, 0), (2, − 2), (− 2, − 4), (0, 0).
(k) (0, 0), (1, − 1), (0, − 4), (− 1, − 4), (0, 0).

2. Calculate the pressure on a semicircular area of 4 ft. radius, assuming the base to lie in the surface of the water.

3. Calculate the pressure on the lower half of an ellipse whose semi-axes are 2 and 3 respectively, (a) when the major axis lies in the surface; (b) when the minor axis lies in the surface.

4. In each of the following parabolas calculate the pressure on that area which lies in the fourth quadrant, assuming the x-axis as the surface of the water:

(a) $x = 4 - y^2$.
(b) $x = (2 + y)(3 - y)$.
(c) $x = (y + 4)(2 - y)$.
(d) $y = x^2 - 4x$

5. A symmetrical parabolic area measures 2 ft. across the base and 1 ft. from base to vertex. With the axis of the parabola perpendicular to the surface of the water, calculate the pressure on the area, (a) when the vertex lies in the surface; (b) when the base lies in the surface.

6. A symmetrical parabolic area measures 12 ft. across the base and 20 ft. from base to vertex. With the axis of the parabola perpendicular to the surface of the water, calculate the pressure on the entire area when so placed (a) that the vertex lies in the surface; (b) that the base lies in the surface.

7. A horizontal cylindrical tank is half full of oil weighing 50 lb. per cubic foot. The diameter of each end is 4 ft. Calculate the pressure on each end. Calculate also the pressure when the tank is full.

8. A flood gate 8 ft. square has its top even with the surface of the water. Find the pressure on each portion into which it is divided by a diagonal. (Assume a cubic foot of water to weigh 62 lb.)

9. Find the total pressure on the bottom and sides of a vertical cylindrical tank of radius 3 ft. and depth 5 ft. when filled with water weighing 62 lb. per cubic foot.

10. A water gate in a vertical dam is 10 ft. long and 4 ft. high. Calculate the total water pressure against it when the crest of the water is 4 ft. above the top of the gate. ($W = 62$ lb. per cubic foot.)

11. If the water gate in example 10 was circular and 8 ft. in diameter, what would be the pressure when the water was 2 ft. above the top of the gate?

12. If the water gate in example 10 was in the form of a semiellipse of base 8 ft. and 3 ft. high (base downwards), calculate the pressure when the water (a) is level with the top of the gate, (b) is 2 ft. above this level.

After studying Exercise CX on the center of gravity the student should return to this exercise and show that the *pressure in each case equals the area of the surface of the gate multiplied by the distance from the surface of the water to the center of area of the gate.*

EXERCISE CIX

Integration as a Process of Summation. Applied Problems. The following list covers a variety of topics, the purpose being to give the student some idea of the wide range of usefulness of this principle. For a more detailed explanation of the theory of some of the problems, he must turn to a text in mechanics or applied mathematics, if it is not contained in the calculus text being used. In all cases, even if the form of the differential element is given, the student should draw a figure and set up the element for himself, as that is the most valuable part of the exercise. Where the formulas derived are general, the student should evaluate them for various constants.

1. A rectangular gate a units long and b units wide is in the face of a dam which makes an angle θ with the horizontal. If the upper edge of the gate is l units vertically below the surface of the water, calculate the pressure on the gate.

2. Solve example 1 for the case where the gate is 8 ft. long and 3 ft. wide and the upper edge 4 ft. below the surface. (Assume water to weigh 62 lb. per cubic foot and $\theta = 60°$.)

3. (a) A cylinder of length l and radius a is submerged so that its axis makes an angle θ with the normal to the surface. The center of the cylinder is c units below the surface. Find the pressure on each end. (b) Evaluate (a) for a 10 ft. cylinder of radius 2 ft. if the center is 4 ft. below the surface. (Take $\theta = 60°$ and assume that water weighs 62 lb. per cubic foot.)

4. Find the total pressure on a sphere 6 ft. in diameter submerged so that its center is 10 ft. below the surface.

5. The sections of the retaining walls on the sides of a canal are in the form of a 30° arc of a circle of radius 16 ft. with the center of the circle on the surface of the water at its normal depth. Calculate the total pressure per linear foot against each side wall. Calculate also the vertical and horizontal pressures. (Take $W = 60$ lb. per cubic foot.)

6. Wind is blowing against a vertical circular tower with a velocity of v ft. per second. The formula for wind pressure against a surface at right angles to its direction of motion is $P = kv^2/g$ lb. per square foot, where k is the weight of a cubic foot of air in pounds. What is the total pressure on the tower if it is h feet high and has a radius of a feet?

Hint. The element of pressure is the pressure against a vertical strip of width ds. The *normal* pressure, N, therefore, against an elementary strip is $P \cos \phi \, ds$; but obviously the resultant pressure at right angles to the wind's direction is zero; hence we need only consider the component of N parallel to the wind's direction, which is $N \cos \phi$.

7. Using the data of example 6, calculate the total pressure exerted by a 15-mi.-per-hour wind against a circular tower 20 ft. in diameter and 100 ft. high. (Assume $k = .0804$ and $g = 32$.)

8. The calculation of work done which is defined as Force times Distance furnishes an example for summation. The work done in raising M pounds a vertical distance of h feet against gravity is defined as Mh foot-pounds. Calculate (a) the work done in pumping out a cylindrical well of radius a and depth h feet; (b) the same for a conical well 50 ft. deep, 4 ft. in diameter at the top, assuming that water weighs $62\frac{1}{2}$ lb. per cubic foot.

9. A horizontal cylindrical oil tank is to be emptied by a pump the outlet of which is 2 ft. above the top of the tank. Calculate the work required. The tank is 40 ft. long and its diameter is 6 ft. (Assume the oil to weigh 60 lb. per cubic foot.)

10. The part of a cistern filled with water is of the form of the solid obtained by revolving $x^2 = 7 - y$ about the y-axis from $y = 3$ to $y = 6$, the positive end of the y-axis being downward. The unit is 1 yd., and the top of the cistern is 9 ft. above the surface of the water. Calculate the work done in pumping it out if water weighs 62 lb. per cubic foot.

11. When the limiting curve of an area has a simple polar equation, this fact can often be used in obtaining an integral over this area. For example, find the pressure against a vertical circular water gate of diameter a feet, the water being level with the top of the gate. (Draw a figure. Take the pole at the highest point of the gate and show that the element of pressure is $dP = k \cdot \rho \sin\theta \cdot \rho d\rho d\theta$. Here our integral is a double integral, but simple.)

12. A horn-shaped solid is formed by rotating a variable circle C about a line in its plane. The nearer extremity of the diameter of C, which is perpendicular to the axis of rotation, traces out a quadrant AB of a circle of radius a, and the radius of C is always $k\theta$, where θ is the angular measure of the variable arc on AB. Find the volume.

13. The area of any curved surface may be found as a double integral. The element of surface is taken as that portion of the surface which projects an element of area on a coördinate plane, say $dA = dx\,dy$ on the XY-plane. This gives $dS = dx\,dy/\cos\gamma$, or, since $\cos\gamma$ is a direction cosine, $S = \iint\sqrt{\left(\frac{\partial z}{\partial x}\right)^2 + \left(\frac{\partial z}{\partial y}\right)^2 + 1}\cdot dx\,dy$. The limits are determined by the projection of the surface on the XY-plane. Calculate the area of a hemisphere by this method.

14. Using the discussion of example 16 as a guide, calculate the area cut from the sphere $x^2 + y^2 + z^2 = 16$ by the cylinder $x^2 + y^2 - 4x = 0$. Calculate also the area cut from the cylindrical surface by the sphere.

Hint. In the second case project on another plane, say the XZ-plane.

15. Find the area of the cylindrical surface $x^2 + y^2 = r^2$ between the planes $y = mx$ and $z = 0$.

16. Calculate the area of the surface of the paraboloid $y^2 + z^2 = 2\,ax$, cut out by the cylinder $y^2 = ax$ and the plane $x = 4\,a$.

17. In problems concerning the torsion of a circular pulley shaft, it is necessary to calculate the moment of the shearing force for a cross section of the shaft. Here the element is the moment of the force on a ring of radius x and infinitesimal width dx. Its area is $dA = 2\,\pi x\,dx$. Call the shearing force F kilograms per square centimeter. Then the moment of force of the shear on this ring is $FdA \cdot x$. Calculate the moment of the shearing force if the radius of the shaft is a centimeters and $F = F_0x/a$. F_0 is the value of F at the periphery of the shaft and is to be regarded as constant.

18. Calculate the moment of the shearing force for a hollow shaft of inner radius b and outer radius a.

19. The results found in examples 17 and 18 measure the permissible turning moment which may be applied to a shaft by a belt or crank, provided merely that F_0 be the maximum allowable value of the unit shearing force in the material of the shaft. The value of F_0 is determined by the elastic limit of the material and is an experimental constant. Which is the stronger, a solid shaft of radius $3a$ or a hollow one of inner radius $3a$ and outer radius $4a$? Note that the solid shaft has the larger area of cross section. What is the relative strength? (The two shafts are assumed to be made of the same material.)

20. The attraction between two particles is proportional to the product of their masses and inversely proportional to the square of the distance between them. Calculate the attraction of a thin bar of length l and mass M for a small particle of mass m which is in the line of the axis of the bar and a units from one end of it.

21. Calculate the attraction of a wire bent in the form of a semicircle AB upon a particle of mass m at O, the center of the circle.

Hint. We need concern ourselves only with the attraction along OP, the radius of the circle which goes through the mid-point of AB, as the resultant at right angles to this is zero.

22. Calculate the attraction of a disk of radius a for a particle of mass m which is located h units vertically above O, the center of the disk.

Hint. Take as the element of the disk a ring of infinitesimal width with O as its center. As in example 21, we consider only the component along the perpendicular.

23. A water reservoir is in the form of a hemisphere 20 ft. in diameter surmounted by a frustum of a cone 8 ft. high and 32 ft. in diameter at the top. Calculate in short tons the total pressure on the walls when the reservoir is filled with water weighing $62\frac{1}{2}$ lb. per cubic foot.

24. The illumination at a point is inversely proportional to the square of the distance from the source of light, and directly proportional to the intensity of the source and the cosine of the angle between the ray of light and the surface illuminated. Find the total illumination on a circular plane area of radius a due to a light of intensity I, fixed at a distance h directly above the center of the disk. (Compare with example 22 on the attraction of a disk.)

25. A circle of radius 30 ft. is surrounded by a walk of width 10 ft. (a) Calculate the relative amount of light on the circle and on the walk when a light is suspended 20 ft. vertically above the center of the circle. (b) What is the ratio of the average intensity over the area to the maximum intensity?

26. In determining the friction between a bearing and a journal in the case of a horizontal shaft, we take as the element of area of the bearing surface, dA, an infinitesimal strip lds or $la\,d\theta$. The element of friction, dF, is then the force of friction on dA. Hence, if l is the length of the journal, a its radius, and μ the coefficient of friction, $dF = p\mu dA$ or $p\mu l a d\theta$, where p is the *normal* pressure per square unit of surface and is in general a function of θ. Here μ may also be a function of θ, θ being the central angle subtended by the arc of contact, measured from the vertical. In integrating, therefore, the limits of θ are from 0 to $\pi/2$ for one half the total friction. Assuming $\mu = .004$ and p as a constant everywhere equal to 10 kg. per square centimeter, calculate the total force of friction F in the case of a journal 20 cm. in diameter and 25 cm. long.

27. Calculate F for the bearing in example 26, assuming $\mu = \mu_0 - k\cos\theta$, where $\mu_0 = .004$ and $k = .001$. (It should be noted that the work done per revolution to overcome friction is $2\pi a \cdot F$.)

28. Assume the same dimensions of bearing as in example 26, but with $p = p_0 \cos\theta$, where $p_0 = 10$ kg. per square centimeter.

29. Solve example 27 for the case where $p = p_0 \cos\theta$, and $p_0 = 10$ kg. per square centimeter.

30. Solve example 26, assuming $p = p_0 \cos^{3/2}\theta$, when $p_0 = 10$ kg. per square centimeter.

NOTE. Use approximate methods of integration, if necessary, in these problems.

31. The load resting on the bearing in problems of the type explained in example 26 is important and is to be found by integration. The element of load (that is, the load resting on the infinitesimal strip $la\,d\theta$) is the vertical component of the pressure on this strip, or $la\,d\theta \cdot p\cos\theta$. Calculate the load resting on the bearing in example 28.

32. Calculate the load resting on the bearing in example 30.

33. Solve example 27, assuming that $p = 10(1 - k\theta)$ kg. per square centimeter if $k = .2$. Calculate also the load in this case. (In each of the examples of this type the student will find it desirable to express the relation between the frictional force F and the load P.)

34. When a vertical shaft revolves with a smooth, flat end turning in a shallow socket, we have a *flat step bearing*. Here the bearing surface is a circular area, with the axis of the shaft passing vertically through its center O. The element of area dA can be taken as a ring of infinitesimal width with O as a center (see example 17). The force of friction dF on this element is $\mu p\, dA$, where p is the pressure per unit area on the step. Finally, the work done per revolution against this friction is $dW = 2\pi x dF$. Calculate the total work per revolution against friction if the radius of the shaft is a, μ is a constant, and the pressure of the shaft is evenly distributed over the whole bearing surface.

35. Suppose the vertical shaft in example 34 is hollow with inner radius 10 cm. and outer radius 20 cm. (a) Calculate W if $\mu = .04$ and p is 2 kg. per square centimeter. (b) Calculate W if $\mu = .04$ and $p = k/x$, where k is a constant. Evaluate the result for $k = 30$.

36. If the bearing surface of the vertical shaft of radius a in example 34 is any surface of revolution, the element of area is no longer $2\pi x\,dx$ but $2\pi x\,ds$, a zone of the bearing surface, where ds must be determined from the generating curve of this surface. The element of work of friction, dW, is then $p dA \cdot \mu \cdot 2\pi x = 4\pi^2 \mu p x^2 ds$. Verify this formula by the aid of a figure. Calculate the general expression for W when the bearing surface is a spherical zone of one base of generating arc α.

NOTE. It is best to express x and ds in terms of R, the radius of the sphere, and θ, the central angle. The limits of θ are then 0 and α. Here p is the normal pressure per square centimeter and is to be regarded as a constant.

37. Calculate the load on the bearing in example 36. (Note that, as in example 31, the load is the sum of the vertical components of the normal pressure. Do not confuse R and a in these examples.)

38. Calculate the work of friction in a conical step bearing, the radius of the shaft being a. (Assume p constant.)

39. Calculate the work of friction when the bearing surface of the step bearing is a paraboloid of revolution with the radius of the shaft 8 cm. and the altitude of the paraboloidal pivot 4 cm. (Assume $p = 30$ kg. per square centimeter and $\mu = .05$.) Calculate also the load on the bearing.

40. In general the question asked in several of the preceding problems, namely, to find the load, is the converse of the problem in practice. We ordinarily know the load P and, assuming a law of distribution, calculate p_0. (a) Assume that a load of 1000 kg. rests on the flat step bearing of a hollow vertical shaft of inner radius 5 cm. and outer radius 8 cm. Assume further as a law of distribution $p = p_0(1 - \frac{1}{10}x)$. Find p_0. (b) Assume that a load of 1000 kg. is supported by a journal, as in example 26. If the bearing is 10 cm. long and the radius of the journal is 8 cm., calculate p_0 if the law of distribution is $p = p_0 \cos\theta$. (c) Solve (b), assuming the law $p = p_0 \cos^{3/2}\theta$.

41. If a bearing of the type explained in example 36 is worn, the pressure is no longer evenly distributed, but tends to obey the law (1) $px = k\cos\theta$, where k is a constant and equal to $p_a a/\cos\alpha$. Here α *is the inclination of the profile to the horizontal at the periphery* and a is the radius of the shaft. In example 36, where the bearing is a spherical one, α is equal to the central angle of the generating arc.

(a) Calculate the work of friction for a spherical bearing on a shaft of radius a if the generating arc is α. (b) Calculate the value of p_a if the load on the bearing is 500 kg., the radius of the shaft is 5 cm., and $\alpha = 45°$, and evaluate the work of friction for this case if $\mu = .04$.

42. In applied mechanics cases arise where the total stress on a support can be found as an integral. A single example will illustrate the class. A lock wall 50 ft. wide contains a rectangular culvert 20 ft. wide whose axis is 20 ft. from the face of the wall. The resultant of all the forces acting above the plane of the culvert is vertical, its magnitude is 400,000 lb., and its action line lies 20 ft. from the face of the wall. Assume that the unit stress developed in the two columns into which the culvert divides the wall increases at a uniform rate from a minimum at the back of the wall to a maximum at the face. Find (a) the total load carried by each column, (b) the unit stresses at the two faces of each column, and (c) the rate of increase r of the unit stress per foot of width of wall.

NOTE. The total stress developed in the two columns must equal the applied force, or 400,000 lb. The sum of the moments of the elementary stresses about any point, say the back or face of the wall, must be equal to the moment of the applied force. The moment of a force about an axis is defined as the force multiplied by the distance from the axis to the point where the force is applied. Hence, if S_0 is the unit stress at the back, the element of stress at a distance x from the back of the wall is $dS = (S_0 + rx)\,dx$. The moment of the stress with respect to the back of the wall is $dM = xdS$.

43. Show that the total attraction exerted by a solid sphere of radius R and density k upon a particle of mass m located at a distance a from the center of the solid sphere is Mm/a^2, where M is the mass of the sphere and $a > R$.

NOTE. This is added as an example where spherical coördinates (see Exercise CVII, 13) may be used to advantage. Observe that the total attraction is along the line of centers.

EXERCISE CX

Center of Gravity of an Area. The standard formulas are

$$\bar{x} = \frac{\iint x\,dy\,dx}{\iint dy\,dx} = \frac{M_y}{A}; \quad \bar{y} = \frac{\iint y\,dy\,dx}{\iint dy\,dx} = \frac{M_x}{A}.$$

Note that we are finding the average value of x and of y *over the area*; that is, (x, y) is any point within the area bounded by the given curves. This idea of the "average value" of a function over an area is important and must not be confused in the present case with the average value of the ordinates or abscissas of the bounding curve. The numerators in the above formulas are called first moments, or *moments of area*, and are designated by M_y and M_x respectively. In the following examples plot the bounding curves and calculate the coördinates of the center of gravity $(\bar{x}, \bar{y})$. Mark that point on the figure. It is suggested that the figure in one or two examples be cut from cardboard. If the work has been done accurately, the piece of cardboard will remain balanced when supported on a pin point at its calculated center of gravity or area. If the center of gravity is not within the bounded area, this check is not feasible.

1. $y^2 = x + 1$, $y + x = 1$.

2. $y^2 = 8x$, $x + y = 6$.

3. $y^2 = 4 - x$, $y^2 = 4 - 4x$.

4. $y^2 = 4x$, $y^2 = 5 - x$.

5. $y = x^2 + x$, $y = 2x^2 - 2$.

6. $y = x^2 - 4x$, $y = 6 - x^2$.

7. $y = 2x - x^2$, $y = 3x^2 - 6x$.

8. $4y = 2x - x^2$, $2y = 3x - x^2$.

9. $3y = 3 - x^2$, $y = 2x - x^2 + 1$.

10. $6y = x^2 - x - 2$, $4y = x^2 - 2x - 3$.

11. $x = y^2 - 4y$. $3x + y^2 = 0$.

12. $y^2 = 4 - x$, $y^2 + 2y = x$.

13. $4y = 3x^2$, $2y^2 = 9x$.

14. $x^2 = 2y^3$, $x^2 = 8y$.

15. $y = x^2 - 4x$, $y^2 = 24x$.

16. $x^2 = 16y$, $x = y^2 - 2y$.

17. $y^2 = 2x$, $y = x - x^2$.

18. $y = x^3 - x$, $y = 3x - x^3$.

19. $y = x^2 + x$, $y^2 = 12(1 + x)$.

20. $4y = x^3$, $x = y - y^2 + 4$.

21. $y = (x - 3)^2$, $y = x(x - 3)^2$.

22. Given the curves $x^2 + y^2 - 4x - 4y = 0$ and $4y = 16 - x^2$, calculate the coördinates of the center of gravity or area of

(a) that part of the circle lying outside the parabola;

(b) that part of the circle lying inside the parabola.

23. Given the circle $x^2 + y^2 = 10x$ and the parabola $9y = x^2 - 12x$, calculate the coördinates of the center of gravity of

(a) that part of the circle lying outside the parabola;
(b) that part of the circle lying inside the parabola.

24. Calculate the coördinates of the center of gravity of the area bounded by $x = 0$, $y = 4$, $xy = 4$, $8y = 14x - 3x^2$.

25. Calculate the coördinates of the center of gravity of the area bounded by $y = 0$, $xy = 4$, $8y = 14x - 3x^2$.

26. Given $x^2 + y^2 = 10$ and $y = 3 \sin \pi x/2$. Find the coördinates of the centers of gravity (a) of the area in the first quadrant bounded by these curves and the y-axis; (b) of the area in the second and third quadrants bounded by these curves and the y-axis.

EXERCISE CXI

Average Value of a Function throughout a Region. The process of finding average values may be extended to other problems and is not limited to finding average values of x and y over an area as in the preceding exercise. The average value of $f(x)$ between the values $x = a$ and $x = b$ is $\int_a^b f(x)\,dx \div (b - a)$. The average value of $f(xy)$ over an area is $\iint f(xy)\,dx\,dy \div A$, the limits on the double integral being the same as in finding the area. The average value may equally well be along a curve, over a curved surface, throughout a volume, or may be an integral without any physical or geometrical interpretation.

1. Find the average value of x^2 from $x = 0$ to $x = 10$.

2. Find the average value of the ordinates of $y^2 = 4x$ from (0, 0) to (4, 4) taken uniformly along the x-axis.

3. What is the average value of xy over a quadrant of the circle $x^2 + y^2 = a^2$? for all points on the arc of the circumference?

4. What is the average value of the abscissas of $y^2 = 4x$, (a) when uniformly distributed along the y-axis; (b) when uniformly distributed along the curve? (The latter is given by $\int x\,ds \div \int ds$. Note the difference in the results, and be able to explain this difference.)

5. What is the average value of the z coördinate of the part of the paraboloidal surface $z = 4 - x^2 - y^2$ which lies above the XY-plane, when the ordinates are uniformly dense over the circular base? (This is *not* the distance to the center of gravity of the solid bounded by the surface and the XY-plane.)

6. What is the average value of $\sin t$ from $t = 0$ to $t = \pi$; that is, throughout a half period?

7. What is the average value of $\sin^2 t$ throughout a half period?

8. Find the average value of ρ^2 over the circle $\rho = \sin\theta$. This gives the square of the *radius of gyration.*

9. Find the average value of ρ^2 over the area bounded by the curve $\rho = \sin 2\theta$

10. Find the average value of the length of the radii vectores of the curve $\rho = \sin 2\theta$. Compare with the result in example 9.

11. Given $v = v_0 + 32t$. Find the average value of v (a) during the first five seconds, starting from rest; (b) during the first five seconds, starting with an initial velocity of 36 ft./sec.; (c) during the first two and one-half seconds, starting from rest.

12. Given $v = \sqrt{v_0^2 + 2gh}$. Find the average value of v during the first one hundred feet, (a) starting from rest; (b) starting with an initial velocity of 60 ft./sec. (In the given formula h is the distance.)

NOTE. The student should observe that in both examples 11, (c), and 12, (a), the distance covered is 100 ft. in $2\frac{1}{2}$ sec., but the average values calculated are different. He should go no further until he can explain clearly this difference.

13. The speed of a sled on an ice pond may be taken to equal $v_0 - \mu gt$. If a boy strikes the ice with an initial velocity of 60 ft. per second, what is his average speed during the first ten seconds? ($\mu = \frac{1}{8}$, $g = 32$.)

14. A crank OA and a connecting rod AB move a slider (attached by a pivot at B) in a straight path in the direction of O, the center of the crank shaft. (a) If the crank turns about O with constant angular velocity ω, what is the average value of the velocity of the slider at B? Does it depend on the length of the connecting rod? (b) Evaluate the result for a crank 2 ft. long making 120 revolutions per minute.

15. In examples 13, 14, and 15, Exercise XCIV, we considered the problem of water flowing through a weir or a water gate. Referring to them, calculate the average flow per second over an open weir, or spillway, 10 ft. wide, while the water is falling regularly from a 4-ft. crest to zero overflow. (Take $\mu = 1$.)

16. Water is escaping through a water gate 8 ft. long and 5 ft. high. What is the average rate of flow from the time the water is 4 ft. above the level of the top of the gate until it drops to that level? How many cubic feet escaped if the time occupied was 15 min.? (Take $\mu = 3/4$.)

17. Finding the center of gravity of a homogeneous solid or of a solid the density of which varies according to some fixed law is an application of average value — the average value of a coördinate throughout a mass, that is, $\int x\,dm \div \int dm$ gives the x coördinate. (The element dm must of course be chosen so that all parts of it are equidistant from the YZ-plane. The limits of the two integrals are the same whether single or triple. Find the center of gravity of a solid homogeneous hemisphere.)

18. Calculate the center of gravity of a solid hemisphere when the density is (a) proportional to the distance from the base; (b) proportional to the square of the distance from the base.

19. Calculate the center of gravity of a homogeneous solid in the form of a paraboloid of revolution, with the radius of its base 6 in. and its altitude 9 in.

20. A solid is in the form of a semiellipsoid. The base has semiaxes 2 ft. and 3 ft. respectively, and its altitude is 4 ft. Find the center of gravity.

21. A rectangle moves from a fixed point O, one side varying as the distance from the point, the other as the square of this distance. The center of the rectangle traces out a straight line OP, perpendicular to the moving plane. When OP is 2 ft., the rectangle is a square of side 3 ft. Find the volume and center of mass of the solid generated.

22. Finding the center of pressure against a dam or water gate is another example of average value. As the water pushes against the gate, the turning effect, or torque, about an arbitrary axis, of the water pressure acting on a horizontal strip is proportional to the pressure on the strip and its distance x from the arbitrary axis. The center of pressure is a point such that if P, the total pressure, were exerted at this point on the gate, the torque action would be unchanged. If the arbitrary axis is taken in the surface of the water, we have, therefore, as the formula for $\bar{x}$, the distance of the center of pressure below the surface, $\bar{x} = \int x\,dP \div \int dP$, where $\int dP$ is the pressure integral found in a previous exercise and x is the distance of the typical element below the surface of the water. Calculate the center of pressure on a rectangular gate 8 ft. long and 4 ft. high when the water is level with the top of the gate.

23. Calculate the center of pressure on the gate in example 22 when the level of the water is 6 ft. above the top of the gate.

24. Find the center of pressure on an equilateral triangular gate in a vertical dam, each side of which is 10 ft., one side of the triangle being in the surface of the water.

25. Find the center of pressure in example 24 if one vertex of the gate is in the surface and the opposite side parallel to the surface.

26. The end of a tank filled with water is parabolic in shape. It is 4 ft. across the top and 4 ft. deep. Locate the center of pressure.

27. A gate in the face of a dam is a square with a diagonal parallel to the surface of the water and 6 ft. below it. If each diagonal is 8 ft., calculate the total pressure and the center of pressure. (Assume $W = 62$.)

28. The end of a trench is in the form of a semicircle of radius 2 ft. surmounted by a 4-ft. square. A gate fits in this and is braced along a horizontal line. How far up should the braces be put, so that the gate will not tend to turn about the line of braces when the trench is full of water? How high up should the braces be if there is only 2 ft. of water in the trench?

29. Find the center of pressure on a circular gate of radius 3 ft. when the top of the gate is 4 ft. below the surface of the water.

30. Calculate by triple integration the center of gravity of one eighth of a sphere.

31. Calculate by triple integration the center of gravity of the solids in Exercise CVII, examples 4, 6, and 7.

32. Find the center of gravity of a thin wire in the form of an arc of the parabola $y^2 = 4x$ from the origin to the positive end of the latus rectum. Find the same for the general case $y^2 = 4ax$.

EXERCISE CXII

Moment of Inertia. Plane Areas. The general formula for the moment of inertia of any object with respect to an axis is $\int r^2 dm$, where r is the common distance of all points of the element dm from the axis. From this follow the formulas for the moment of inertia of a plane area with respect to the x-axis, the y-axis, and the origin. They are

$$I_x = \iint y^2 dx\,dy, \quad I_y = \iint x^2 dx\,dy, \quad I_0 = \iint (x^2 + y^2)\,dx\,dy = I_x + I_y.$$

The limits are precisely the same as if we were finding the area by double integration. If we wish to lay stress on the physical significance of moment of inertia, we must speak of the moment of inertia of thin plates, or laminæ, of the shape defined by the area. In that case each formula must be multiplied by δ, the mass of the plate per square unit of area.

Find the moment of inertia for the areas below. The curves bounding the areas are given.

1. $y^2 = 8x$, $y = 2x$.
2. $y^2 = x$, $x + y = 2$, $y = 0$.
3. $y = 4 - x^2$, $y = 0$.
4. $2y = x^3$, $2x - y = 0$.
5. $4y = x^2$, $y^2 = 4x$.
6. $4y = x^3$, $2y = 4x - x^2$.

7. Given the area in the first quadrant bounded by the two curves $y^2 = 2x^3$ and $y^2 = 8x$, find (a) the area, (b) the coördinates of the center of gravity, (c) the moment of inertia with respect to each axis and the origin.

8. Given the area bounded by the parabola $y^2 = x + 1$ and the straight line $x + y = 1$. Solve as in (7).

9. If the moment of inertia is divided by $\int dx\,dy$ (or generally $\int r^2 dm \div \int dm$), we have the average of the squares of the distance of the "particles" from the axis. *The square root of this average value is called the radius of gyration.* This is an important quantity, and the student should find the radius of gyration for each of the preceding examples and for those in the next exercise. Further examples will be met in the study of mechanics, where the importance of both moment of inertia and radius of gyration will be realized.

EXERCISE CXIII

Polar Moment of Inertia. In this case the general integral $\int r^2 dm$ gives $I = \int\int \rho^3\, d\rho\, d\theta$. The areas for which the student is asked to calculate the moment of inertia in this exercise include only such as are bounded by curves whose equations are given in polar coördinates, which system leads to the formula above. The examples are introduced at this point rather as an application of double integration than for their mechanical importance. The more important moments of inertia are included under Routh's rules in the study of mechanics.

Calculate I for that part of the area indicated in each of the following examples :

1. The circle $\rho = 2$, which lies to the right of the line $\rho = \sec\theta$.

2. The circle $\rho = \cos\theta$, which lies outside the circle $\rho = 1/\sqrt{2}$.

3. The circle $\rho = \cos\theta$, which lies in the first quadrant outside the circle $\rho = \sin\theta$.

4. The circle $\rho = \sin\theta$, which lies outside the cardioid $\rho = 1 - \cos\theta$.

5. The circle $\rho = \cos\theta$, which lies outside the cardioid $\rho = 1 - \cos\theta$.

6. The lemniscate $\rho^2 = 2\cos 2\theta$, which lies outside the circle $\rho = 1$.

7. Calculate I for that smaller part of the lemniscate $\rho^2 = 9\cos 2\theta$, which lies outside the circle $\rho = \sqrt{6}\cos\theta$.

8. Calculate I for the area between the two circles $\rho = 3\sin\theta$ and $\rho = 2\sin\theta$

CHAPTER XVIII

DIFFERENTIAL EQUATIONS

EXERCISE CXIV

Differential Equations of the First Order. This list includes only the simplest type of differential equations of the first order, where the variables are separable.

1. $xy\,dx + \sqrt{1 - x^2}\,dy = 0.$
2. $(x + xy^2)\,dy - 3\,dx = 0.$
3. $\sqrt{1 - y^2}\,dx = \sqrt{1 + x^2}\,dy.$
4. $\sqrt{1 + y^2}\,dx - (1 - x^2)\,dy = 0.$
5. $\sin y\,dx + e^x dy = 0.$
6. $(1 + x^2)\,dy - \sqrt{1 - y^2}\,dx = 0.$
7. $(x^2 - yx^2)\,dy + (y^2 + xy^2)\,dx = 0.$
8. $(1 + y^2)\,x\,dx + (1 + x^2)\,dy = 0.$
9. $yx\,dy - (1 - y^2)\,dx = 0.$
10. $y\,(4 + 9\,x^2)\,dy - dx = 0.$
11. $y\,(4\,x^2 - 9)\,dy - 6\,dx = 0.$
12. $(x^2 + 1)\,dy + x(y - 1)\,dx = 0.$
13. $(1 + y^2)\,x\,dx + (1 + x^2)\,dy = 0.$
14. $(2\,x + 1)\,dy + y^2 dx = 0.$
15. $(1 + 2\,y)\,x\,dx + (1 + x^2)\,dy = 0.$
16. $(1 + x)\,dy - (1 - x)\,dx = 0.$
17. $(1 + y^2)\,dy - y\,dx = 0.$
18. $ye^{2x} dx - (1 + e^{2x})\,dy = 0.$

EXERCISE CXV

The Linear Differential Equation. In applied mathematics the student meets the linear differential equation first in the study of mechanics, where it appears as a linear relation involving the displacement x (or s), the velocity dx/dt, and the acceleration d^2x/dt^2. Hence in the following examples it seems advisable to change the independent variable from x to t, thereby making the student more familiar with the notation of the equations with which he will be concerned later. The general linear homogeneous equation is of the form

$$(1)\quad A\frac{d^n x}{dt^n} + B\frac{d^{n-1}x}{dt^{n-1}} + \cdots N\frac{dx}{dt} + Px = 0.$$

The equation (2) $Ar^n + Br^{n-1} + \cdots Nr + P = 0$ is called the *auxiliary equation* of (1). Solve (2), getting n real or imaginary values of r, distinct or equal. Suppose the distinct real roots are $a_1, a_2 \cdots$, the double real roots $b_1, b_1, b_2, b_2 \cdots$ and the imaginary pairs $c_1 \pm d_1 i$, $c_2 \pm d_2 i \cdots$. Then the solution of (1) is

$$A = C_1 e^{a_1 t} + C_2 e^{a_2 t} + \cdots C_k e^{b_1 t} + C_{k+1} t e^{b_1 t}$$
$$+ \cdots e^{c_1 t}(C_r \cos d_1 t + C_{r+1} \sin d_1 t) + \cdots,$$

the groups on the right corresponding to two real roots, a double root, and a pair of imaginary roots respectively. Observe the equation of the second order, $A\frac{d^2x}{dt^2} + B\frac{dx}{dt} + Cx = 0$, particularly. The auxiliary equation is $Ar^2 + Br + C = 0$.

(1) If the roots are real and distinct, say a and b, the general solution is $x = C_1 e^{at} + C_2 e^{bt}$.

(2) If the roots are equal, say a and a, $x = C_1 e^{at} + C_2 t e^{at}$.

(3) If the roots are imaginary, say $a \pm bi$,

$$x = e^{at}(C_1 \sin bt + C_2 \cos bt).$$

Solve the following and verify the results:

1. $\frac{d^2x}{dt^2} - 5\frac{dx}{dt} + 6x = 0.$
2. $\frac{d^2x}{dt^2} + \frac{dx}{dt} - 6x = 0.$
3. $\frac{d^2x}{dt^2} - 4\frac{dx}{dt} + 4x = 0.$
4. $\frac{d^2x}{dt^2} - 4x = 0.$
5. $\frac{d^2x}{dt^2} + 4x = 0.$
6. $\frac{d^2x}{dt^2} + 4\frac{dx}{dt} = 0.$
7. $\frac{d^2x}{dt^2} - 2\frac{dx}{dt} + x = 0.$
8. $\frac{d^2x}{dt^2} - 2\frac{dx}{dt} - x = 0.$
9. $\frac{d^2x}{dt^2} - 4\frac{dx}{dt} + 13x = 0.$
10. $\frac{d^2x}{dt^2} + 4\frac{dx}{dt} + 8x = 0.$
11. $\frac{d^2x}{dt^2} - 6\frac{dx}{dt} + 4x = 0.$
12. $\frac{d^2x}{dt^2} - 6\frac{dx}{dt} + 12x = 0.$
13. $\frac{d^2x}{dt^2} + 3\frac{dx}{dt} + 3x = 0.$
14. $4\frac{d^2x}{dt^2} + 4\frac{dx}{dt} + 5x = 0.$
15. $9\frac{d^2x}{dt^2} - 6\frac{dx}{dt} - 8x = 0.$
16. $\frac{d^2y}{dt^2} - 8\frac{dy}{dt} + 25y = 0.$
17. $\frac{d^2y}{dt^2} + 12\frac{dy}{dt} + 11y = 0.$
18. $9\frac{d^2y}{dt^2} + 12\frac{dy}{dt} - 21y = 0.$

19. $9\frac{d^2y}{dt^2} - 6\frac{dy}{dt} - 5y = 0.$

20. $16\frac{d^2y}{dt^2} + 8\frac{dy}{dt} + 9y = 0.$

21. $9\frac{d^2y}{dt^2} - 3\frac{dy}{dt} - y = 0.$

22. $3\frac{d^2y}{dt^2} + 4\frac{dy}{dt} + 5y = 0.$

23. $\frac{d^3x}{dt^3} + \frac{d^2x}{dt^2} - 12\frac{dx}{dt} = 0.$

24. $\frac{d^4x}{dt^4} - 4\frac{d^2x}{dt^2} = 0.$

25. $\frac{d^4s}{dt^4} + 3\frac{d^2s}{dt^2} - 4s = 0.$

26. $\frac{d^4s}{dt^4} - 5\frac{d^2s}{dt^2} + 4s = 0.$

27. $\frac{d^4s}{dt^4} + 5\frac{d^2s}{dt^2} + 4s = 0.$

28. $\frac{d^3s}{dt^3} + 3\frac{d^2s}{dt^2} + 3\frac{ds}{dt} + s = 0.$

29. $\frac{d^4s}{dt^4} + 3\frac{d^3s}{dt^3} + 3\frac{d^2s}{dt^2} + \frac{ds}{dt} = 0.$

30. $\frac{d^5s}{dt^5} - 4\frac{ds}{dt} = 0.$

EXERCISE CXVI

The Linear Differential Equation: Second Order. The case to be studied is

$$A\frac{d^2x}{dt^2} + B\frac{dx}{dt} + Cx = f(t).$$

In this exercise we are concerned with the solution of the linear differential equation of the second order only, whose right-hand member is not zero but a function f of the independent variable t. The complete solution has the form $x = u + v$, where u, *the complementary function*, is the solution of the left-hand member set equal to zero (see preceding exercise), which already contains the maximum number of arbitrary constants, and where v, *the particular solution*, is assumed with undetermined coefficients, which are determined, however, by demanding that v shall satisfy the original equation, of which it must be a solution.

The form of v, which we assume with undetermined coefficients, depends on the nature of the function $f(t)$ and the relation of f to u. We shall tabulate three cases with which the student should be familiar:

(A) when $f(t)$ is a polynomial;
(B) when $f(t)$ is an exponential function;
(C) when $f(t)$ is a trigonometric function.

(A) *When $f(t)$ is a Polynomial of Degree n.* Assume v as the most general polynomial of this degree, and determine the unknown coefficients by demanding that v be a solution of the given differential equation. For example, if $f(t)$ is of the second degree, assume $v = at^2 + bt + c$. Substitute this value of v for x in the left-hand side of the original equation. Equate coefficients of like powers of t, and solve for a, b, and c. Solve the following:

1. $\frac{d^2x}{dt^2} - 6\frac{dx}{dt} + 13x = 39.$

2. $\frac{d^2x}{dt^2} - 4\frac{dx}{dt} + 7x = 14.$

3. $\frac{d^2x}{dt^2} - 2\frac{dx}{dt} - 3x = 2t + 1.$

4. $\frac{d^2y}{dt^2} + 9y = t + \frac{1}{2}.$

5. $\frac{d^2y}{dt^2} + 9y = t^2.$

6. $\frac{d^2x}{dt^2} - 2\frac{dx}{dt} + 4x = t^2.$

7. $\frac{d^2x}{dt^2} - 6\frac{dx}{dt} + 12x = 6t^2.$

8. $\frac{d^2y}{dt^2} - 10\frac{dy}{dt} + 25y = 25t^2 + 5t + 17.$

9. $\frac{d^2s}{dt^2} + 4\frac{ds}{dt} + 13s = 13t^2 + 21t + 19.$

10. $\frac{d^2y}{dt^2} - 2\frac{dy}{dt} + 5y = 5t^2 - 4t + 17.$

11. $\frac{d^2x}{dt^2} + \frac{dx}{dt} - 2x = -2(1 + t + t^2).$

12. $\frac{d^2y}{dt^2} - 6\frac{dy}{dt} - 6y = -6t^2 - 14t - 18.$

13. $\frac{d^2y}{dt^2} - 4y = -4t^2 - 8t - 10.$

14. $\frac{d^2x}{dt^2} - \frac{dx}{dt} - 12x = 36 - 26t - 12t^2.$

15. $\frac{d^2x}{dt^2} + 2\frac{dx}{dt} + 8x = 4t^2 - 6t - 3.$

16. $\frac{d^2x}{dt^2} + 4x = 4t^3 + 8t^2 + 18t + 20.$

(B) *When $f(t)$ has the Form e^{nt}.* If e^{nt} does not occur in u, assume $v = ae^{nt}$ and determine a as before. If e^{nt} does occur in u, assume $v = ate^{nt}$ and determine a as before. Apply these methods to the solution of the following:

17. $\frac{d^2y}{dt^2} - \frac{dy}{dt} - 2y = e^{2t}.$

18. $\frac{d^2y}{dt^2} - 3\frac{dy}{dt} - 4y = e^{-t}.$

19. $\frac{d^2y}{dt^2} - 2\frac{dy}{dt} - 3y = e^{2t}.$

20. $\frac{d^2x}{dt^2} - 2\frac{dx}{dt} - 3x = e^{3t}.$

21. $\frac{d^2x}{dt^2} - 2\frac{dx}{dt} - 3x = e^{-t}.$

22. $\frac{d^2x}{dt^2} - 9x = e^{3t}.$

23. $\frac{d^2x}{dt^2} + 9x = 9e^{3t}$.

24. $\frac{d^2s}{dt^2} - \frac{ds}{dt} - 2s = 8e^{3t}$.

25. $\frac{d^2s}{dt^2} - \frac{ds}{dt} - 2s = -8e^{-t}$.

26. $\frac{d^2x}{dt^2} - \frac{dx}{dt} - 6x = 5e^{3t}$.

27. $\frac{d^2x}{dt^2} - 2\frac{dx}{dt} + x = 6e^{t}$.

(C) *When f(t) has the form k sin nt (or k cos nt).* If the first derivative is present in the differential equation of the second order, always assume $v = a \sin nt + b \cos nt$ and determine a and b as before. This is also always true if $f(t) = k_1 \sin nt + k_2 \cos nt$.

If the first derivative is missing, that is, $B = 0$ in the equation at the beginning of this exercise, then $\sin nt$ and $\cos nt$ may occur separately (that is, not combined with an exponential) in u. If they do so occur in u, **assume $v = at \cos nt$ when $f = k \sin nt$; $v = at \sin nt$ when $f = k \cos nt$.** If they do *not* so occur in u, **assume $v = a \sin nt$ when $f = k \sin nt$; $v = a \cos nt$ when $f = k \cos nt$.** Apply this method to the solution of the following:

28. $\frac{d^2x}{dt^2} + 9x = 5\cos 2t$.

29. $\frac{d^2x}{dt^2} + 9x = 3\cos 3t$.

30. $\frac{d^2x}{dt^2} - 9x = 6\cos 3t$.

31. $\frac{d^2x}{dt^2} + 4x = 10\sin 3t$.

32. $\frac{d^2x}{dt^2} + 4x = 8\sin 2t$.

33. $4\frac{d^2x}{dt^2} + x = 4\sin t/2$.

34. $2\frac{d^2y}{dt^2} - y = \sin t$.

35. $\frac{d^2s}{dt^2} + 4s = 8\cos 2t$.

36. $\frac{d^2s}{dt^2} + 9s = -6\sin 3t$.

37. $\frac{d^2s}{dt^2} - 16s = 2\cos 4t$.

38. $\frac{d^2y}{dt^2} - 2\frac{dy}{dt} - 3y = 8\cos 2t$.

39. $\frac{d^2x}{dt^2} + 6\frac{dx}{dt} + 13x = 30\sin t$.

40. $\frac{d^2x}{dt^2} - 2\frac{dx}{dt} + 5x = 10\sin t$.

41. $\frac{d^2x}{dt^2} - 2\frac{dx}{dt} + 5x = 17\sin 2t$.

EXERCISE CXVII

Miscellaneous Differential Equations. In this list of equations the student will meet with some new forms together with an opportunity of reviewing the types already studied and of combining the various principles used in the previous section. If we find equations of the second order like those in the preceding exercise except that $f(t)$ is a sum of two or more types A, B, and C, v is the sum of the v's we would obtain if only one type of terms was present in the right-hand member.

1. $\frac{d^2x}{dt^2} = t^2.$

2. $\frac{d^2x}{dt^2} = 4 \sin 2t.$

3. $\frac{d^2x}{dt^2} = x.$

4. $\frac{d^2x}{dt^2} = e^{2t}.$

5. $\frac{d^2x}{dt^2} - 4x = 5e^{3t} - 3e^t.$

6. $\frac{dy}{dx} + \frac{y^2}{2x+1} = 0.$

7. $\frac{d^2y}{dt^2} + 4y = 5 \sin 3t - 10 \cos 3t.$

8. $\frac{d^2y}{dt^2} + 4y = 4e^{2t} + 4t^2.$

9. $\frac{d^2y}{dt^2} + 4y = 16 - 5 \sin 3t.$

10. $\frac{d^2y}{dt^2} + 4y = 8e^{2t} + 15 \sin t/2.$

11. $8\left(\frac{dy}{dt}\right)^3 = 27y.$

12. $\frac{d^4y}{dt^4} - 6\frac{d^3y}{dt^3} + 13\frac{d^2y}{dt^2} = 0.$

13. $\frac{d^4y}{dt^4} - 4\frac{d^3y}{dt^3} + \frac{d^2y}{dt^2} = 0.$

14. $4\frac{d^2s}{dt^2} + s = 4 \sin t/2.$

15. $4\frac{d^2y}{dt^2} - 2\frac{dy}{dt} + 7y = \sin t/2.$

16. $\frac{d^2y}{dt^2} - 2\frac{dy}{dt} + y = 2e^{3t} + 3t^2 - 7.$

17. $4\left(\frac{dy}{dx}\right)^2 = 9x.$

18. $\frac{d^2y}{dt^2} - \frac{dy}{dt} - 6y = 6e^{4t} - 8e^{-t}.$

19. $\frac{d^2x}{dt^2} - 2\frac{dx}{dt} - 3x = 4e^{3t} - 12.$

20. $\frac{d^2x}{dt^2} - 2\frac{dx}{dt} + 5x = 5e^{2t} + 11e^{-t}$

21. $\frac{d^2s}{dt^2} + 4\frac{ds}{dt} + 13s = 4 \cos 3t - 12 \sin 3t.$

22. $\left(\frac{dy}{dx}\right)^3 - 27y^2 = 0.$

23. $\frac{d^2x}{dt^2} + \frac{dx}{dt} - 2x = 4e^{2t} + 2t + 3.$

24. $\frac{d^2x}{dt^2} - \frac{dx}{dt} - 6x = -5e^{-2t} - 6t - 18.$

25. $\frac{d^2x}{dt^2} - 2\frac{dx}{dt} + 5x = 8 \sin t - 4 \cos t + 5t - 2.$

26. $\frac{d^2s}{dt^2} + \frac{ds}{dt} - 6s = 6t + \sin t.$

27. $\frac{d^2y}{dt^2} + 25y = 10\cos 5t.$

28. $\frac{d^2x}{dt^2} - 2\frac{dx}{dt} + x = 2e^t.$

29. $\frac{d^2y}{dt^2} - 3y = 2te^t.$

30. $\frac{d^2y}{dt^2} - 4y = \sin t + 3te^t.$

In the next four equations set $y = vx$, after which the variables can be separated. This method can be applied to all homogeneous equations of the first order.

31. $ydx + (x + y)dy = 0.$

32. $x \sin(y/x)dy - y\sin(y/x)dx + x\,dx = 0.$

33. $(x^2 + y^2)dx - 2xy\,dy = 0.$

34. $2x^2dx + (x^2 + y^2)dy = 0.$

In the next two equations multiply both sides by $2\frac{ds}{dt} \cdot dt$ before integrating the first time. Two integrations will be required to get the solution. This method can be applied to many equations of the form $d^2s/dt^2 = f(s)$ with advantage.

35. $\frac{d^2s}{dt^2} = \frac{4}{s^3}.$

36. $\frac{d^2s}{dt^2} = \frac{1}{(s+1)^3}.$

The next four examples are differential equations of the first order of the form $\frac{ds}{dt} + Ps = Q$, where P and Q are functions of t. Use the substitution $s = uv$ where $u = e^{-\int Pdt}$. The substitutions are given, but should be verified.

37. $\frac{ds}{dt} + \frac{s}{t} = t.$ Let $s = v/t.$

38. $\frac{ds}{dt} + s\tan t = \tan t.$ Let $s = v\cos t.$

39. $\frac{ds}{dt} + \frac{2t}{t^2+1}s = \frac{1}{t}.$ Let $s = v/(t^2+1).$

40. $\frac{ds}{dt} + 2ts = t^3.$ Let $s = ve^{-t^2}.$

Many differential equations of the first order can be solved by a properly chosen transformation not included in previous types. For a complete discussion, see a treatise on differential equations.

41. $t^2\dfrac{ds}{dt} - 2\,st - s^3 = 0.$ Let $s = t^2/v.$

42. $(t^2 + t)\,ds = (t^2 + 2\,st + s)\,dt.$ Let $s = vt.$

43. $(3 + 2\,st)\,s\,dt + (3 - 2\,st)\,t\,ds.$ Let $st = v.$

44. $(x + y)^2\dfrac{dy}{dx} = 2\,x + 2\,y + 5.$ Let $x + y = v.$

EXERCISE CXVIII

Application of Differential Equations. Evaluating the Constants. We have already evaluated the constant of integration in Exercise LXXVII, where the examples were, in fact, linear differential equations of the first order. In the case of differential equations of the second order there are two general constants to be evaluated. The simplest case is just the reverse of finding the second derivative. In the problems below, the student should know that if s represents distance, d^2s/dt^2 is f, the acceleration. From mechanics he should know that force = mass × acceleration, or $F = mf$. If g appears in a problem without explanation, it is to be taken as 32 ft./sec.^{2}

1. Given $\dfrac{d^2s}{dt^2} + 2\dfrac{ds}{dt} + 5\,s = 0$, with the initial conditions $s = 4$, $v = 0$, when $t = 0$. Express s in terms of t.

2. Given $\dfrac{d^2s}{dt^2} + 4\,s = 8\,t^2$, with the initial conditions $s = 0$, $v = 0$, when $t = 0$. Express s in terms of t.

3. Given $\dfrac{d^2s}{dt^2} + 4\,s = 3\sin t$, with the initial conditions $s = 0$, $v = 5$, when $t = 0$. Express s in terms of t.

4. Given $\dfrac{d^2s}{dt^2} + 4\,s = 4\cos 2\,t$, with the initial conditions $s = 2$, $v = 2$, when $t = 0$. Express s in terms of t.

5. The acceleration of a falling body is 32 ft./sec.^{2} (a) If a body is projected downward with an initial velocity of 24 ft./sec., find the relation between s and t. (This relation is called the *equation of motion*.) (b) If the body is 72 ft. above the ground when projected, how long will it take it to reach the earth's surface?

Hint. In (a) $s = 0$, $v = 24$, when $t = 0$.

6. The acceleration of a body projected vertically upward is -32 ft./sec.[2] If projected with a velocity of 96 ft./sec., find the equation of motion and determine how high the body will rise.

7. The motion of a *simple pendulum* is defined by the law $d^2\theta/dt^2 = -g \sin\theta/l$. For small displacements we take $\sin\theta = \theta$. Assuming this, find the equation of motion if $\theta = \alpha$, and $d\theta/dt$, or ω, the angular velocity, is zero at the start. Take the pendulum 2 ft. long. Find also the time required for a complete swing of the pendulum.

8. The force acting on a unit particle is proportional to the distance of the particle from the starting point and is directed towards the starting point (that is, force $= -ks$). The magnitude of the force is 16 poundals (32 poundals $= 1$ lb.) when $s = 1$ ft. What is the equation of motion of a particle projected away from the origin with a velocity of 16 ft./sec.? This is called an *undamped, or harmonic, vibration.*

9. In the case of a *damped vibration* there is another force (friction or air resistance) which is proportional to the speed and opposing the motion. Hence the general equation is $m\dfrac{d^2s}{dt^2} = -k_1\dfrac{ds}{dt} - k_2 s$. Calculate the equations of motion for a unit mass starting away from the origin with a velocity of 16 ft./sec., assuming $k_1 = 4$ and $k_2 = 8$.

10. The *discharge of an electric current* from a condenser of low resistance is defined by the law $d^2C/dt^2 = -k^2C$, where k is a constant. If $k = 500$, find the current C in terms of t, the time. What is the frequency?

NOTE. When any motion or change of state is periodic, the frequency is the reciprocal of the period.

11. A certain *flywheel being stopped by friction* obeys the law $d^2\theta/dt^2 = -4$. Derive a general expression for θ, the angular distance in radians, in terms of t. Assume that ω, the angular velocity, is 240 revolutions per minute at the start. When will the wheel come to rest? How many revolutions will it have made?

Hint. $\omega = d\theta/dt$ and should be expressed in radians per second to determine the constants of integration. θ will be in radians.

12. A constant force acts on a flywheel so that it obeys the law $d^2\theta/dt^2 = 2$. Express θ in terms of t, assuming the flywheel to start from rest. How long will it take for it to acquire a speed of 1000 revolutions per minute? How many revolutions will it make in the first minute?

13. If a *particle of mass m is vibrating under the force of a spring* and loses energy, owing to friction, at a rate proportional to the square of its velocity, it obeys the law $\dfrac{d^2s}{dt^2} + \dfrac{k}{m}\dfrac{ds}{dt} + \dfrac{a}{m}s = 0$, where a is a constant

and k is the factor of proportionality, depending on the spring. Find the equation of motion for a unit mass. Evaluate this for the case $k = \frac{1}{2}$, $a = 1\frac{5}{8}$, assuming the spring was displaced (stretched) 4 units at the start. (This important law holds for fluid and electromagnetic friction, and is equivalent to saying the frictional force is proportional to the velocity.)

14. A *condenser* of capacity K, discharging through a circuit of resistance R, obeys the law $\frac{d^2V}{dt^2} + \frac{R}{L}\frac{dV}{dt} + \frac{1}{LK}V = 0$. In this equation L is the self-induction and V the potential. Express V in terms of t, assuming $V = V_0$, $dV/dt = 0$, when $t = 0$.

15. An equation similar to the one in example 14 arises in the theory of the *ballistic galvanometer*. The equation here is $I\frac{d^2\theta}{dt^2} + k_1\frac{d\theta}{dt} + k_2\theta = 0$, where I is the moment of inertia of the needle, k_1 is a constant determined by the oil bath or other device used to retard the motion, k_2 is a constant depending on the torsion of the fiber and the magnetic field, and θ is the angular displacement. (a) Derive the general expression for θ, assuming $4\,Ik_2 > k_1^2$. (b) Evaluate the constants of integration, assuming a displacement α from its equilibrium position when the needle is released.

16. Suppose $k_1^2 > 4\,Ik_2$ in example 15. What is peculiar about the motion in this case?

17. Given, as in example 14, a condenser of capacity K, self-induction L, and resistance R in the circuit. Apply to this an electromotive force of V volts and the law giving Q, the charge, is $L\frac{d^2Q}{dt^2} + R\frac{dQ}{dt} + \frac{Q}{K} = V$. (a) Solve this equation for Q, assuming V to be constant. (b) What form does this assume if $Q = 0$, $dQ/dt = 0$, when $t = 0$?

18. (a) Solve the equation of example 17, assuming that V varies with the time, that is, $V = kt$. (b) Assume V periodic in example 17, or $V = k \sin t$, and solve.

19. The problem of a beam supported in different ways and with a variously distributed load furnishes a series of very simple differential equations, a few of which follow. Calculate in the following cases the *curve of mean fiber* (that is, the relation between y and x) under the given conditions. In cases (a) and (b) find the deflection y of the middle of the beam. In all cases locate the points of inflection.

(a) A uniform beam of length l loaded at one end and rigidly fixed at the other. The differential equation is $d^2y/dx^2 = k(l - x)$, with $y = 0$, $dy/dx = 0$, when $x = 0$.

(b) A uniform beam as in (a) with load uniformly distributed along the beam. The differential equation is $d^2y/dx^2 = k(l^2 - 2\,lx + x^2)$, with $y = 0$, $dy/dx = 0$, when $x = 0$.

(c) Uniform beam with load uniformly distributed and supported at both ends. The differential equation is $d^2y/dx^2 = k(l^2 - 4x^2)$, with $y = 0$, $dy/dx = 0$, when $x = 0$. (Here the origin is at the middle of the beam and y is the vertical height of any point above the middle of the beam.)

(d) If the beam in (c) is rigidly embedded at both ends, the equation is $d^2y/dx^2 = k(\frac{1}{24}l^2 - \frac{1}{2}x^2)$, with $y = 0$, $dy/dx = 0$, when $x = 0$. Show that the points of inflection in this case are nearer the center than in (c).

(e) A beam of uniform strength with thickness equal to $1/(a + bx)$, rigidly fixed at one end. The differential equation is $d^2y/dx^2 = k(a + bx)$, with $y = 0$, $dy/dx = 0$, when $x = 0$.

20. The general differential equation of the curve of mean fiber of a loaded beam (cf. example 19) is $M = EI\, d^2y/dx^2$, where $M =$ the bending moment of a section x units from the end, $E =$ the coefficient of longitudinal elasticity, and $I =$ the moment of inertia of the section x. For a beam fixed at both ends and uniformly loaded with W pounds per linear inch, the bending moment at any section is given by the equation $M = Wlx/2 - Wx^2/2 - Wl^2/12$. (a) Deduce the general equation for the curve of mean fiber, and calculate its maximum and minimum points and its points of inflection. (b) Evaluate this result for a rectangular beam of wood 8×12 in. and 24 ft. long, carrying a load of 24 lb. per linear foot. I for a rectangular cross section is $bd^3/12$, where d is the dimension perpendicular to the neutral axis of the cross section. E for wood is 1,000,000. Calculate the amount of the maximum deflection. (c) Consider a steel beam 2×3 in. E for steel is 30,000,000.

21. The discharge in cubic feet per second over an open weir is $Q = cbH^{3/2}$, where $c = 3.5$, $b =$ width of the weir, $H =$ depth of overflow. (a) A lake covering 164 sq. mi. is controlled by an overflow weir, or spillway, 2000 ft. wide. Assume a flood in the basin drained by the lake which is discharging 12,500 cu. ft. per second into the lake at the time water begins to overflow. The flood discharge increases to 182,800 cu. ft. per second in 29 hr. How long must this maximum discharge be continued to cause the overflow to rise 5 ft. on the weir crest? (b) Suppose the water did not begin to flow over the crest until the discharge reached 73,500 cu. ft. per second, how long would it take the water to rise 5 ft. on the crest? (c) Suppose the flood should come at a time when the river had been at the freshet stage long enough to fill the lake and give a free flow of 12,500 cu. ft. per second (corresponding to a depth of about 1.5 ft. over the crest), how long would it require the maximum flow to be continued after it was reached to give a depth of 5 ft. on the crest?

NOTE. H may be assumed to increase uniformly until the flood discharge reaches its maximum.

ANSWERS

(In the answers the constant of integration in simple indefinite integrals is omitted.)

EXERCISE I (PAGE 1)

13. $f(1) = 1, f(2) = 1.301, f(100) = 3.$
14. $f(1) = 64, f(4) = 2\sqrt{2},$ $f(11) = 1.4595.$
15. $f(1/2) = 11.28, f(2) = 37.62.$
16. $f(0) = 2, f(1) = 9.5234.$
17. $f(2) = 2.198, f(3) = 4.240.$
18. $f(1) = 2.624, f(2) = -1.838.$
19. $f(1) = 13.868, f(4) = -55.935.$

EXERCISE II (PAGE 2)

1. $2/3.$ **2.** $1/3.$ **3.** $-1/2.$ **4.** $\frac{1}{2}.$ **5.** $0.$ **6.** $\infty.$ **7.** $3/5.$ **8.** $-5.$ **15.** $1.$ **16.** $1.$

EXERCISE III (PAGE 4)

13. $-1/x^2.$
14. $-1/3t^2.$
15. $-2/x^3.$
16. $-3/(x+1)^2.$
17. $2/(1-t)^2.$
25. $-1/(3-x)^2.$
26. $-2(x+1)/x^3.$
27. $2x(x+1)/(2x+1)^2.$
31. $\tan^{-1} 4/3.$
32, 33. $\tan^{-1} 3$ and $\tan^{-1} 3/19.$
34. $\tan^{-1} 8/15.$
38. $\pi/2.$

EXERCISE IV (PAGE 5)

1. $x^{-\frac{1}{2}} - 2x^{-\frac{1}{3}}.$
2. $3\sqrt{x} + 1/\sqrt{2x}.$
6. $3x^2 - \frac{1}{9}x^{-\frac{4}{3}}.$
11. $\sqrt{3}/2\sqrt{x} + 1/6\sqrt{x}.$
16. $-8/(x-1)^3.$
17. $x/\sqrt{x^2-4}.$
18. $-x^2/(1-x^3)^{2/3}.$
19. $9/(1-x)^4.$
20. $-3x/\sqrt{4-3x^2}.$
21. $-3x/\sqrt{x^2-2}.$

EXERCISE VI (PAGE 6)

1. $2(x^2+4x+2)/\sqrt{x^2+4x}.$
2. $(4x^2+5x-6)/\sqrt{x^2-3}.$
3. $(12t^2-4t-3)/\sqrt{1-2t^2}.$
4. $(15t^3-6t^2-18t+4)/\sqrt{2t^3-4t}.$
5. $(8x^2-19x+5)/\sqrt{1-2x+4x^2}.$
10. $(2x^2+6x+2)/(1-x^2)^2.$
11. $(4x^2+4x+3)/(1+2x)^2.$
12. $(3t^2-2t)/(3t-1)^2.$
13. $(4t^3+3t^2-8)/(1+2t)^2.$
17. $-5/(x^2-5)^{3/2}.$
18. $(4-3x)/2x^3\sqrt{x-1}.$
19. $-(x^3-2x+1)/(1-x^2)^{3/2}.$
25. $(3+2x-6x^2)/\sqrt{(2x^2-3)(1-2x)}.$

EXERCISE VII (Page 7)

1. $3x^2 - 4x + 1/x^{3/2}$.
2. $1 - 1/x^{3/2}$.
3. $3/2\,x^{3/2} + 1/\sqrt{3} + 1/2\sqrt{3x}$.
5. $(3x^2+2)/\sqrt{2x^3+4x-6}$.
6. $(3x^2-1)/2x^2\sqrt{3x+1/x}$.
8. $(2x^2-4x+3)/\sqrt{x^2-2x+2}$.
15. $3t/\sqrt{t-2}$.
16. $x/\sqrt{2x+4}$.
17. $(1-8t)/\sqrt{1-4t}$.
18. $-t/(1+t)^{3/2}$.

EXERCISE VIII (Page 8)

1. $4(x+2)/(x^2+4x+6)$.
2. $6(6x+1)/(6x^2+2x+5)$.
4. $-x/(1-x^2)$.
18. $1/\sqrt{1+x^2}$.
19. $2/\sqrt{4x^2+9}$
20. $1/\sqrt{9x^2-1}$.
28. $(x^{3/2}+2)/2x(x^{3/2}-1)$.
29. $-2/\sqrt{x^2+1}$.
30. $(x^4+3x^2-2)/x(x^4-1)$.

EXERCISE IX (Page 9)

1. $-\log_7 e \cdot 3x^2/(1-2x^3)$.
2. $\log_{10} e/(3x+7)$.
3. $\log_4 e \cdot \left(2\sqrt{6x}+3\right)/\left(2\sqrt{6}\,x^{3/2}+6x\right)$.
4. $-13\log_4 e/(x-3)(5-6x)$.
5. $\log_8 e \cdot (x^2-1)/(x^3+x)$.
6. $\log_a e \cdot (2x^2+1)/(x^3+x)$.
7. $-\log_5 e \cdot 4t/(t^4-1)$.
8. $\log_2 e \cdot 2x/(1-x^4)$.
9. $\log_2 e \cdot (2-3x^2)/(x-x^3)$.
10. $\log_6 e \cdot (4t^3-t^2-2)/(1-2t)(1-t^3)$.

EXERCISE X (Page 10)

1. $4(3x^2-2)/2x^3-4x \cdot \log(2x^3-4x)$.
2. $(x+3)/(x^2+6x)\sqrt{\log(x^2+6x)}$.
3. $3(2x+7)/(x^2+7x)\cdot\log^2(x^2+7x)$.
4. $2(2x^3+3x)/(x^4+3x^2)\cdot\log\sqrt{x^4+3x^2}$.
5. $-15x/(1-3x^2)\cdot\log^4\sqrt{1-3x^2}$.
6. $-6x/(1-2x^2)\sqrt{\log(1-2x^2)}$.

EXERCISE XI (Page 11)

1. $2xe^{x^2+2}$.
2. $e^{\log_3 x}\cdot\log_3 e/x$.
3. $e^{\sqrt{x}}/2\sqrt{x}$.
4. $xe^{\sqrt{x^2+4}}/\sqrt{x^2+4}$.
5. $(2x+1)\,e^{\log(x^2+x)}/x^2+x$.
6. $-xe^{\sqrt{1-x^2}}/\sqrt{1-x^2}$.
14. $(3t+2)\,e^{t\sqrt{1+t}}/2\sqrt{1+t}$.
15. $2x\log_4 e\cdot e^{\log_4(x^2+4)}/(x^2+4)$.
16. $(x-1)\,e^{(x+1)/\sqrt{x}}/2x^{3/2}$.
17. $e^{1/\sqrt{1-x}}/2(1-x)^{3/2}$.
21. $(-x+2x\log 1/x)\,e^{x^2\log 1/x}$.
22. $\left(1-1/2\sqrt{x}\right)e^{x+1/\sqrt{x}}$.

EXERCISE XII (Page 11)

1. $2xa^{x^2} \cdot \log a$.

2. $(2t+2)2^{t^2+2t} \cdot \log 2$.

3. $(1+\log t)5^{t\log t} \cdot \log 5$.

4. $2^{\sqrt{x}}\log 2/2\sqrt{x}$.

5. $-7^{-1/x} \cdot \log 7/x^2$.

16. $2/t \cdot \log t/2 \cdot c^{\log^2 t/2}\log c$.

17. $xa^{\sqrt{x^2+5}/3}\log a/3\sqrt{x^2+5}$.

18. $-3xa^{3/\sqrt{x^2+5}}\log a/(x^2+5)^{3/2}$.

EXERCISE XIII (Page 12)

1. $2x(x^4+5x^2+7)/(x^2+1)^{3/2}(x^2+7)^{1/2}$.

2. $(2x^4+3x^2-3)/x^4(1-x^2)^{3/2}$.

3. $5x^7+29x^5+40x^3/(x^2+4)^{1/2}(x^3+3x)^{2/3}$.

4. $-(75x^3-14x^2-55x+6)(2x-5)^2/(1-5x^2)^2(x^2-1)^{3/2}$

5. $3x(3x^3+7x^2+25x+70)/\sqrt{2x+7}(x^2+5)^{3/2}$.

20. $-35x(2x-1)/(3x^2-4x^3)^{1/6}$

21. $-\frac{2}{3}(1+17x)\sqrt{1-x}\sqrt[3]{1+x}$.

EXERCISE XIV (Page 12)

1. $x^{1/x}(1-\log x)/x^2$.

2. $(1-x)^{x^2}(2x\log(1-x)-x^2/(1-x))$.

3. $(1+x)^{x^2}(2x\log(1+x)+x^2/(1+x))$.

4. $x^{\sqrt{3x}}\sqrt{3/x}(2+\log x)$.

5. $3(t^2-6t)^t[\log(t^2-6t)-(2t-6)/(t-6)]$.

6. $3x^{x^2+3}(2x\log x+(x^2+3)/x)$.

19. $(x\log x)^x[(1+\log x)/\log x+\log(x\log x)]$.

20. $(x+1)^{e^x}[e^x/(x+1)+e^x\log(x+1)]$.

21. $x^{e^{x+1}}(e^{x+1}/x+e^{x+1}\log x)$.

22. $(x^2+4)^{x/4}[\frac{1}{4}\log(x^2+4)+x^2/2(x^2+4)]$.

EXERCISE XV (Page 13)

1. $3x^2\cos x^3$.

2. $2\cos(2x+5)$.

3. $-2/x^2 \cdot \cos 2/x$.

4. $1/2\sqrt{x} \cdot \cos\sqrt{x}$.

5. $2xe^{x^2}\cos e^{x^2}$.

6. $2\sin(3-2x)$.

7. $-2e^{2t}\sin e^{2t}$.

11. $4x\sec^2(2x^2+3)$.

12. $e^x\sec e^x\tan e^x$.

13. $2/x^3 \cdot \csc^2 1/x^2$.

14. $-2x\csc(x^2+1)\cot(x^2+1)$.

15. $-1/t^2 \cdot e^{1/t}\sec^2 e^{1/t}$.

19. $-2x\cos x^2 \cdot \sin(\sin x^2)$.

20. $-2(e^x-e^{-x})\sin(2e^x+2e^{-x})$.

EXERCISE XVI (PAGE 14)

1. $6x \sin^2 x^2 \cos x^2$.
2. $-1/\sqrt{x} \cdot \cos\sqrt{x} \sin\sqrt{x}$.
3. $-4/x^2 \cdot \tan^3 1/x \sec^2 1/x$.
4. $5/x \cdot \sec^5 \log x \tan \log x$.
5. $-6e^{2t} \cot^2 e^{2t} \csc^2 e^{2t}$.
6. $-24e^{-3t} \sin^3 e^{-3t} \cos e^{-3t}$.
7. $6 \tan(x+3) \sec^2(x+3)$.
8. $2/x^2 \cos^2 1/3x \sin 1/3x$.
9. $4 \cos 2x/\sqrt{\sin 2x}$.
10. $-6x^2 \sin x^3/\cos^{2/3} x^3$.
20. $4/(1-x)^2 \sin^3 x/(1-x) \cdot \cos x/(1-x)$.
21. $-8t/(t^2-1)^2 \cdot \sec^2 1/(t^2-1) \cdot \tan 1/(t^2-1)$.
24. $e^y(y+1)\sin(2ye^y)$.
25. $2\cos 2x \sin(2\sin 2x + 2)$.

EXERCISE XVII (PAGE 15)

1. $3\cos 3t \cos 2t - 2\sin 3t \sin 2t$.
2. $\frac{1}{2}e^{2t}\cos t/2 + 2e^{2t}\sin t/2$.
3. $-2/t^2 \cos 2/t\, e^{\sin t/2}$.
4. $\frac{1}{2}(\sin t/2 \sec^2 t/2 + \sin t/2)$.
5. $x^{\sin x}(\cos x \log x + \sin x/x)$.
6. $2a^{\sec x} \cdot \log a \cdot \sec x \tan x$.
7. $2te^{t/2}\sec^2 t^2 + \frac{1}{2}e^{t/2}\tan t^2$.
10. $6 \cot 3\theta$.
11. $\tan t/2$.
12. $(\sin t)^t \cdot (\log \sin t + t \cot t)$.
13. $3 \sec 3x$.
19. $-1/2x^2 \cdot \sec 1/2x \csc 1/2x$.

EXERCISE XVIII (PAGE 15)

1. $2/\sqrt{1-4x^2}$.
2. $-2x/\sqrt{1-x^4}$.
3. $3/(1+9x^2)$.
4. $3x^2/x^3\sqrt{x^6-1}$.
5. $-1/x\sqrt{x^2-1}$.
6. $-e^x/(1+e^{2x})$.
7. $-2/(x^2+4)$.
8. $-2x/x^2\sqrt{4x^4-1}$.
9. $1/x\sqrt{4x^2-1}$.
10. $1/2\sqrt{x-x^2}$.
11. $-1/(1-x)\sqrt{x^2-2x}$.
12. $2xe^{x^2}/\sqrt{1-e^{2x^2}}$.
20. $3/t(9+\log^2 t)$.
21. $2/(x^2+1)$.
22. $2/(x^2+1)$.
23. $\cos x/(1+\sin^2 x)$.

EXERCISE XIX (PAGE 16)

1. $-2x/\sqrt{1-4x^2} + \text{arc} \cos 2x$.
2. $e^{2x}/\sqrt{1-e^{2x}} + e^x \text{ arc} \sin e^x$.
3. $2/(4+x^2) \text{ arc} \tan x/2$.
4. $(5t+12)/(t^3+4t)$.
5. $\text{arc} \sin x/2$.
6. $((x^2-4)/(x^2+4))^2$.
7. $26(x-1)/(x^3-6x^2+13x)$.
8. $(t+2)/t\sqrt{t-4}$.
9. $x^2 \text{ arc} \tan x$.
18. $(x^2-3)/x^4 \cdot \text{arc} \sin 1/x$.

EXERCISE XX (PAGE 17)

1. $3\sqrt{1+\log x}/2x$.
2. $3\sqrt{x^2-1}/x^4$.
4. $-x \sin x^2 \cos\sqrt{\cos x^2}/\sqrt{\cos x^2}$.
5. $2(\text{arc} \sin \dfrac{x}{a} + 1)/\sqrt{a^2-x^2}$.
6. $-28/(4x+3)^2$.
8. $12/(x^2+4x)$.
9. $2(3e^{4x}-5)/(3e^{4x}+5)$.
12. $e^x\sqrt{e^x-1}/(e^x+3)$.
13. $\tan x(2+\sec x)$.
16. 1.

EXERCISE XXI (Page 17)

1. $-(2x+y)/(x+2y)$.
2. $2xy/(8y-x^2)$.
3. $(6xy+4)/(2-3x^2)$.
4. $(2x-3y)/(3x-3y^2)$.
5. $-2x(y+1)/(x^2+3)$.
6. $(3x^2+2y)/(2y-2x)$.
15. $(e^{x-y}+e^{x+y})/(e^{x-y}-e^{x+y})$.
16. $\cot x \cot y$.
17. $(2x^2-y^2)/xy(2\log xy+1)$.
18. $e^{2x}(y^2+4)/y$.

EXERCISE XXII (Page 18)

1. $2xe^{x^2}\log e+\sqrt{3}/2\sqrt{x}-1/x^3+e^{\sqrt{x}}/2\sqrt{x}$.
2. $16/(4+x^2)^2$.
3. $x^2/\sqrt{x^2-1}$.
4. $1/t^4\sqrt{t^2+1}$.
5. $1/2\sqrt{x^2+4x}$.
8. $-(2xy+y^2-4)/(x^2+2xy+4)$.
9. $e^{x/2}/(1-e^{x/2})^2$.
10. $\sqrt{8x-x^2}$.
11. $\log(4x^2-1)$.
12. $\log(4x^2+1)$.
13. $\text{arc tan}\sqrt{x}$.
14. $t^2/\sqrt{4-t^2}$.
15. $1/t^8\sqrt{t^2-4}$.
16. $-\text{arc sin}\, x/2\sqrt{1-x}$.
17. $2x\csc(1-x^2)\sec(1-x^2)$.
18. $\sec^3 x$.
19. $4x^2\sqrt{x^2+4}$.
21. $2\sin 1/x/x^2e^{1/x}$.
33. $10x/9(2-x^{1/2})^{4/3}$.

EXERCISE XXIII (Page 20)

1. $1, \sqrt{3}/2, \sqrt{2}/2, 0, -1$.
2. (a) $(3, 3), (3, 1)$; (b) $(4, 2), (2, 2)$.
3. $(3, -4)$; $(-3, 4)$.
4. $(3, -5)$; $(-3, 13)$.
7. When $x=4$.
8. $(4, 4)$.
12. (a) $(1, 2), (-1, 0)$; (b) $(2, 9), (-2, -7)$.
18. (a) $((8n+1)\pi/4, \sqrt{2}), ((8n+5)\pi/4, -\sqrt{2})$;
    (b) $(2n\pi, 1), (\frac{1}{2}(4n+3)\pi, -1)$.
21. (a) $(-2, -4), (-2, 0)$; (b) $(0, -6), (0, -4)$ and $(\frac{1}{2}, -\frac{21}{4}), (\frac{1}{2}, -\frac{35}{8})$.
22. (a) $\pi/4$; (b) 0.
31. $(2, 2)$.
34. $2y=x^2$.
35. $y=\sin x$.

EXERCISE XXIV (Page 22)

1. $18^\circ.4$.
2. $71^\circ.6$.
7. At $(4, 5)$, $9^\circ.5$; at $(-1, 5/4)$, $33^\circ.7$.
8. At $(2, 2)$, 31°.
9. At $(0, 0)$, 24°; at $(8, 2)$, $69^\circ.7$.
10. $26^\circ.6$.
22. At $(2, 2)$, $18^\circ.4$.
23. $x+\sqrt{2}y=6$, $\sqrt{2}x-y=0$, $s_t=4$, $s_n=2$.
24. At $(2, -2)$, $3x+5y+4=0$, $5x-3y-16=0$, $s_t=\frac{10}{3}$, $s_n=\frac{6}{5}$;
    at $(2, 8)$, $18x-5y+4=0$, $5x+18y-154=0$, $s_t=\frac{20}{9}$, $s_n=\frac{144}{5}$.
28. $x+6y+10=0$, $6x-y-14=0$, $s_t=12$, $s_n=\frac{1}{3}$.

EXERCISE XXV (Page 23)

1. $3x - y - 5 = 0,\ x + 3y - 15 = 0.$
2. $x - 4y + 15 = 0,\ 4x + y - 25 = 0.$
3. $x + 3y + 1 = 0,\ 3x - y - 7 = 0.$
4. $3x + 2y + 1 = 0,\ 2x - 3y - 8 = 0.$
7. $x + 6y - 12 = 0,\ 6x - y - 35 = 0.$
8. $4x + 2y - 3 = 0,\ 2x - 4y + 1 = 0.$
9. $x - y = 2\sqrt{3},\ x + y = 10\sqrt{3}/3.$
11. $\sqrt{2}x - y - 1 = 0,\ x + \sqrt{2}y - 2\sqrt{2} = 0.$
12. $15x - 4y - 81 = 0,\ 4x + 15y + 123 = 0.$

EXERCISE XXVI (Page 24)

4. $\pi/3.$
5. $\pi/6.$
6. $\pi/3.$
7. 0 and $\tan^{-1}3\sqrt{3} = 79°.1.$
8. 0 and $\tan^{-1}\frac{3}{5}\sqrt{3} = 46°.1.$
11. $\pi/3.$
14. High points, $\theta = 3\pi/4$ and $9\pi/4$; low points, $\theta = 0$ and $3\pi/2$; extreme left (vertical tangents), $\theta = 3\pi/8$ and $15\pi/8$; extreme right, $\theta = 9\pi/8$ and $21\pi/8$.
15. High, $\theta = 3\pi/8$; low, $\theta = 7\pi/8$; left, $\theta = 5\pi/8$; right, $\theta = \pi/8$.
18. $\tan^{-1}(-9/13)$, or about 145°

EXERCISE XXVII (Page 25)

1. $s = 33$; $v = 12$; $f = -18$; $v = 0,\ t = 2.$
2. $s = 116$; $v = 72$; $f = 6$; $v = 0,\ t = 6.$
3. $s = 72$; $v = 15$; $f = -12$; $v = 0,\ t = 4.$
9. $s = 1$; $v = 3$; $f = 15$; never at rest.
13. $x = -.330$; $v = -.926$; $f = 1.522$; $t = .760$ and $2.331.$
14. $x = .994$; $v = -2.543$; $f = 3.097$; $t = 2.356$ and $5.698.$
20. $x = -3.90$; $v = .975$; $f = -9.379$; $t = 1.899$ and $3.899.$
21. $x = -2.5749$; $v = 2.277$; $f = 1.158$; $t = 3.84$ and $7.84.$
25. 36 sec. 288 ft./sec.
26. The second goes 48 ft. the higher.
27. 1/2 sec., 80 ft./sec.
28. 1824 ft.

EXERCISE XXVIII (Page 27)

1. 2, 2, 1, − 3.
2. 1, 1, 2/3, 2/3.
3. − 3/2, − 3/2, − 1/2.
4. 1/3, 1/3, − 1/2.
17. (a), (d), (e), (f), (i), (j), have double roots.

EXERCISE XXIX (Page 28)

1. $y'' = -\sin x - \cos x,\ y''' = -\cos x + \sin x,$
$y^{(n)} = \sin(x + n\pi/2) + \cos(x + n\pi/2).$
2. $y'' = 2/x^3,\ y''' = -6/x^4,\ y^{(n)} = (-1)^{n-1}\cdot \underline{|n}\cdot 1/x^{n+1}.$
3. $y^{(n)} = 2^n e^{2x}.$
9. $y'' = 2/(1 + x^2)^2,\ y''' = -8x/(1 + x^2)^3.$
10. $y'' = -2e^{-t}\cos t,\ y''' = 2e^{-t}(\sin t + \cos t).$
20. $y'' = \cos t - t\sin t,\ y''' = 2\sin t - t\cos t.$
27. $y'' = (2 - x^2)/(1 - x^2)^{3/2}.$

EXERCISE XXX (Page 28)

1. $(2xy + 3y^3)/(x - 3y^2)^3$.
2. $d^2y/dx^2 = 36/(2y - 3x)^3$, $d^3y/dx^3 = -972x/(2y - 3x)^5$.
3. $-2/(4y - 2x)^3$.
8. $-(\sin x \sin y + \cos^2 x \cot y)$.
9. $-4e^x/e^{2y}$.
10. $-\cos(x + y)/\sin^3(x + y)$.
11. $\sin(x + y)/\cos^3(x + y)$.
12. $(1 - \log x)/x^2$, $d^3y/dx^3 = (2\log x - 3)/x^3$.

EXERCISE XXXI (Page 29)

1. $t/2$.
2. $(3t^2 + 2)/4t^3$.
3. $2/3t^3$.
4. $6(3t^3 + 2t)/(6t + 2)^3$.
5. $1/t$.
6. $1/2\cos^3 2t$.
7. $-b/a^2 \sin^3 t$.
17. $-\csc^2 t/(t \sin t)$.
19. $-(1 + 2\cos t)^3/2(2 + \cos t)^3$.

EXERCISE XXXII (Page 30)

1. Max., $(-1, 1)$; min., $(4, -124)$; inf., $(3/2, -61\frac{1}{2})$.
2. Max., $(-2, 69)$; min., $(3, -56)$; inf., $(\frac{1}{2}, 6\frac{1}{2})$.
3. Max., $(-2, 23\frac{1}{3})$; min., $(6, -62)$; inf., $(2, -19\frac{1}{3})$.
4. Max., $(-6, -50)$; min., $(-3, -82)$; inf., $(-4, -66)$.
5. Max., $(-1, 45)$; min., $(3, -51)$; inf., $(1, -3)$.
9. Max., $(2, 4)$; min., $(1, 3)$ and $(3, 3)$.
17. Max., $(1, 3)$; min., $(-1, -3)$; inf., $(0, 0)$.
18. Max., $(-2, -4)$; min., $(2, 4)$.
38. Max., $(2, 5/3)$; min., $(0, -1)$. 39. Min., $(-3, 13\frac{1}{2})$; inf., $(0, 0)$.

EXERCISE XXXIII (Page 31)

1. Max., $(\pi/2, 1)$; min., $(3\pi/2, -1)$; inf., $(0, 0)$, $(\pi, 0)$.
2. Max., $(\pi/4, 1.553)$; min., $(5\pi/4, -40.944)$; inf., $(0, 1)$.
3. Max., $(2, 3.515)$; min., $(-2, -3.515)$.
7. Max., $(.253, 1.125)$; min., $(1.571, 0)$; inf., $(.883, .580)$.
8. Max., $(.635, 1.76)$; min., $(2.446, -1.751)$; inf., $(1.571, 0)$.
13. Max., $(.551, .517)$; min., $(2.122, -.107)$; inf., $(1.107, .265)$.
14. Max., $(.766, 1.639)$; min., $(2.286, -1.123)$; inf., $(1.447, .343)$.
16. Min., $(1.448, -2.351)$; max., $(3.02, 1.071)$.

EXERCISE XXXIV (Page 32)

1. 18×24 ft.
2. 3×4 ft.
3. 12.54 ft.
4. 4 and 4.
5. 3 in.
6. 18×24 rd.
7. $6 \times 6 \times 10$ ft.
8. 2 ft.
9. $R = 10$ ft.; $h = 30$ ft.
10. 40 ft. wide at top, 16 ft. at the bottom, 16 ft. deep.
11. 2 wk. hence.
12. 150.
13. $46\frac{2}{3}$ rd. from A.
14. $R = 4$.
15. 3 mi. from D.
16. 5 mi. from A.
17. $R = 4$.
18. 1250.
19. $e = 3.51$, $h = 2.23$.

20. Rectangle, 8×6 ft.

21. Parallelopiped, $2 \times 2 \times \frac{1}{2}$ ft.

22. Parallelopiped, $2 \times 2 \times 1\frac{1}{2}$ ft.

23. (a) $4\sqrt{3}$; (b) 6.31; (c) 8; (d) 24.

24. 75,000 @ $7.50 per M.

25. 2 hr. later; 16π cu. ft. increase.

26. 9 mo.

27. 6 mi.; 32 min. later.

28. 2 hr. later; 54.08 mi.

30. Edge of cube = diameter of sphere.

31. $R = 13.37$, $h = 8.91$.

32. Each 5 in.

33. 4 in.

34. 10×5 ft.

35. (a) $x = 2a$, $y = 2b$; (b) $x = a + \sqrt[3]{ab^2}$, $y = b + \sqrt[3]{a^2b}$;
(c) $x = a + \sqrt{ab}$, $y = b + \sqrt{ab}$; (d) $x = a + b^2/a$, $y = b + a^2/b$.

36. $(2, \pm 4)$.

37. $a = b = 5$ in.

38. Each $\sqrt{\frac{1}{2}ab}$.

39. $12\sqrt{3}$ ft.

40. $S = 5.237$, $S_A = 5.250$.

41. 2.813.

42. 3.215.

43. 7.871.

EXERCISE XXXV (Page 37)

1. About 55°.

2. Each is $2\sqrt{10}$.

4. $(-1, -3)$.

5. $e = 2$ in. $h = 2/3\sqrt{3}$.

6. $BP = DA/\sqrt{2}$.

11. 6 in.

15. 4 in. and $5\frac{1}{3}$ in.

16. (a) $r : h = 1 : \sqrt{2}$.
(b) $r : h = 1 : 2\sqrt{2}$.

17. 40 ft.

19. 20 wk.

20. $a^{3/2}c/(a^{3/2} + b^{3/2})$ units from A.

21. 40 mi./hr.

22. 6 mi./hr.

25. 35 mi., 16°.2.

27. $l = 13\sqrt{13}$, $h = 26$.

28. 19.81 ft.

30. $20\sqrt{2}$ ft.

34. $\theta = \pi/4 + \alpha/2$.

35. $\theta = \alpha + \beta$.

37. $\sqrt{2aW/n}$.

41. $x = 12.97$.

46. $\alpha = \frac{1}{2}(90 - \phi)$.

49. $4\frac{4}{5}$ ft.

50. 6.152 in.

51. Parabola: $4ay^2 = b^2(4x - a)$.

54. $CF = \sqrt{ab}$.

55. $\left(l + l\sqrt{1 + 8\sec^2\alpha}\right)/4\sec^2\alpha$; (a) 51.23 ft.

60. $a\sqrt[3]{I_1}/\left(\sqrt[3]{I_1} + \sqrt[3]{I_2}\right)$ from S_1.

EXERCISE XXXVI (Page 45)

1. $ds = r\,dx/\sqrt{r^2 - x^2} = r\,dy/\sqrt{r^2 - y^2}$, $s = 1.014$.

2. $ds = \sqrt{(a + x)/x}\,dx = dy/\left(2a\sqrt{4a^2 + y^2}\right)$, $s = 1.783$.

3. $ds = \sqrt{4a^2x^2 + 4abx + b^2 + 1}\,dx$, $s = 2.429$.

4. $ds = (x^4 + 1)\,dx/2x^2$, $s = 3.879$.

17. $ds = \sqrt{1 + 4t^2}\,dt$.

18. $ds = t\sqrt{4 + 9t^2}\,dt$.

19. $ds = 4\,dt$.

20. $ds = 2\cos t\sqrt{1 + 4\tan^2 t}\,dt$.

25. (17) $\sin\tau = 2t/\sqrt{1 + 4t^2}$, $\cos\tau = 1/\sqrt{1 + 4t^2}$;
(18) $\sin\tau = 3t/\sqrt{4 + 9t^2}$, $\cos\tau = 2/\sqrt{4 + 9t^2}$.

26. (a) $v_x = .97$, $v_y = -3.88$; (b) $v_x = 1.789$, $v_y = -3.578$;
(c) $v_x = 4$, $v_y = 0$.

32. (a) $v_x = 32\sqrt{2}$, $v_y = 16\sqrt{2}$; (b) $v_x = 32\sqrt{2}$, $v_y = 0$;
(c) $v_x = 32\sqrt{2}$, $v_y = \pm 16\sqrt{2}$.

EXERCISE XXXVII (PAGE 47)

1. $ds = a\,d\theta$, $\sin\psi = \cos\theta$, $\cos\psi = -\sin\theta$.
2. $ds = b\,d\theta$, $\sin\psi = \sin\theta$, $\cos\psi = \cos\theta$.
3. $ds = 5\,d\theta$, $\sin\psi = \frac{1}{5}(4\sin\theta + 3\cos\theta)$, $\cos\psi = \frac{1}{5}(4\cos\theta - 3\sin\theta)$.
5. $ds = \sqrt{2 + 2\cos\theta}\,d\theta$, $\sin\psi = \sqrt{(1 + \cos\theta)/2}$,
   $\cos\psi = -\sin\theta/\sqrt{2 + 2\cos\theta}$.
12. $ds = 18/\sqrt{1 + \cos\theta} \cdot d\theta$.
13. $ds = 3/(2 - \cos\theta)\sqrt{5 - 4\cos\theta}\,d\theta$.
17. $ds = \sec^3\theta/2\,d\theta$.
18. $\sqrt{5 - 4\sin 3\theta + 8\cos^2 3\theta}\,d\theta$.

EXERCISE XXXVIII (PAGE 47)

1. $dy = \frac{1}{2}(\cos x + \cos x/2)\,dx$.
2. $(-2/3\,x^{4/3} + 2/x^{1/3} + 5/x^2)\,dx$.
3. $(4x - 5x^2)/2\sqrt{2 - x} \cdot dx$.
4. $5\left(x^2 - \sqrt{x}\right)^4/\left(2x - 1/2\sqrt{x}\right) \cdot dx$.
5. $(2 + x)/(4x + x^2)^{\frac{1}{2}} \cdot dx$.
6. $-3x^2/2\sqrt{1 - x^3} \cdot dx$.
12. $a\sin(2a/x)/x^2 \cdot dx$.
13. $-1/\left(x\sqrt{1 - x^2}\right) \cdot dx$.
19. $e^x(x - 3)/x^4 \cdot dx$.
23. $-1/\left(4\sqrt{x} - 2x\right) \cdot dx$.
26. $-6x\cos^2 x^2 \sin x^2\,dx$.

EXERCISE XXXIX (PAGE 48)

1. $x\,d^2x/dy^2 - y(dx/dy)^2 = 0$.
2. $1 + (dx/dy)^2 - (y - 2)\,d^2x/dy^2$.
5. $(d^2x/dy^2) + (dx/dy)^2 + y - 4 = 0$.
9. $xz\,d^2z/dx^2 + z\,dz/dx + z^2 = 0$.
10. $d^2z/dx^2 + 2\,dz/dx - 2\cos z = 0$.
11. $d^2z/dx^2 - 2\,dz/dx = 0$.
12. $\tan z\,d^2z/dx^2 + (dz/dx)^2 = 0$.
15. $d^2u/dy^2 + u = 0$.
16. $t\,d^2y/dt^2 + dy/dt + y = 0$.
17. $d^2u/dx^2 + 4u = 0$.
19. $d^2y/dz^2 + y = 0$.
25. $d^2y/dt^2 + y = 0$.
26. $(d^2y/dt^2)^2 + (dy/dt)^2 + 1 = 0$.

EXERCISE XL (PAGE 50)

1. 144π ft.^{2}/sec.
2. $8\frac{1}{3}$ ft./sec.
3. 6 ft./sec.
4. $dV/dt = 4\sqrt{3}$ ft.^{3}/hr., $dS/dt = 15.46$ ft.^{2}/hr.
5. $v = 8$ ft./sec., $f = 9/10$ ft./sec.^{2}
6. 6 ft./sec.
7. $25\pi/12$ ft.^{3}/min.
8. 3.717 ft.^{3}/min.
9. $16/75\pi$ ft./sec.
10. $4/3\pi$ ft./min.
13. (a) 8 ft.^{3}/sec.; (b) in 2/3 sec.
15. $3\frac{5}{7}$ ft./sec.
18. (a) 13.823 ft./hr.; (b) 9.840 ft./hr.
21. 7 lb. per sq. in. per sec.
22. 3.27 mi./hr.
25. $14/\sqrt{41}$ in./sec.
26. (a) 1 in./min.; (b) increasing 8 in.^{2}/min.
27. (a) $50\pi/3$ ft./sec.
29. (a) 8/13 ft./sec.; (b) $4\sqrt{3}/\sqrt{109} = .621$ ft./sec.
31. $-16/75$ units per sec. $f = 32/375$ units per sec.^{2}
33. 8.66 ft./sec. and 9.45 ft./sec.
37. 3.1134 ft./sec.
40. $2\frac{1}{2}$ units per sec.
41. 5π ft.^{2}/sec.
42. .731 ft.^{3}/sec.
43. 6.12 in./sec.; 6.16 in./sec.
47. 1/4.
48. Decreasing $\frac{1}{20}$.

EXERCISE XLI (Page 57)

1. 3/2. **2.** $-1/4$. **3.** 1/6. **4.** 1/2, 3/5. **5.** 3/20. **13.** $-1/3$. **14.** 8/69. **15.** 296/9. **16.** ± 1. **17.** $3a$, 0. **21.** $2r$.

EXERCISE XLII (Page 58)

1. 4. **2.** 3. **3.** 0. **4.** 3. **5.** 1/2. **6.** ∞. **9.** $\pi^2/2$. **16.** 0. **17.** 0. **18.** -2.

EXERCISE XLIII (Page 59)

1. 0. **2.** 0. **3.** 1. **4.** $4/\pi$. **5.** $2/\pi$. **6.** 0. **11.** 1/3. **17.** π. **18.** 1. **19.** 1. **20.** ∞.

EXERCISE XLIV (Page 59)

1. 1. **2.** 1. **3.** e^2. **4.** 1. **5.** 1. **6.** $e^{2/\pi}$. **7.** 1. **16.** $1/e$. **17.** e. **18.** $1/e$.

EXERCISE XLV (Page 61)

1. $R = 2\sqrt{2}$, N is $2x + 2y = 3$.
2. $5\sqrt{10}/3$, $x + 3y = 1$.
3. $1/2$, $x = 2$.
4. $8\sqrt{2}/3$, $x - y = 3$.
5. $29\sqrt{29}/25$, $2x + 5y = 12$.
6. $4\sqrt{2}$, $x - y + 1 = 0$.
7. $73\sqrt{73}/100$, $8x + 3y - 22 = 0$.
12. $5\sqrt{5}/4$, $2x + y + 2 = 0$.
13. $125/12$, $3x + 4y - 13 = 0$.
14. $5\sqrt{5}/2$, $2x - y - 6 = 0$.
19. $39/8$, $\sqrt{3}x + y - 3\sqrt{3} = 0$.
21. $R = 2\sqrt{2}$.
22. $R = 1/2$.
23. $R = 1/36$.
24. $R = 5\sqrt{5}$.
28. $R = (x^2 + y^2)^{3/2}/a^2$.
30. Max., $x = n\pi$; min., $x = (2n + 1)\pi/2$.
33. Max. at vertex.
34. Max. at (1/4, 1/4), $R = 2\sqrt{2}$; $R = 2$.

EXERCISE XLVI (Page 62)

1. (3, 3/2). **2.** $(3, -8\frac{3}{4})$. **3.** $(8/3, -5/6)$. **4.** $(21\frac{1}{6}, 15\frac{9}{20})$. **5.** $(-2, 3)$. **6.** $(\pi/2, 0)$. **7.** $(-17\frac{1}{2}, 3\frac{1}{2})$. **8.** (2, 5).
17. $(-7, 8)$. **18.** (7/4, 7/4).

EXERCISE XLVII (Page 63)

1. $a = -8t^3$, $b = 6t^2$. 2. $a = t - \frac{1}{4}t^5$, $b = \frac{5}{6}t^3 + 2/t$.
3. $a = 3 - t + \frac{9}{4}t^5$, $b = 2/3t + 3t^2 + \frac{3}{2}t^5$.
4. $a = 1 + t - \frac{1}{4}t^5$, $b = 2/t + \frac{5}{6}t^3$.
5. $a = (12t^4 + 1)/2t^3$, $b = (4t^4 + 3)/t$.
6. $a = -32t^3$, $b = (7 - 24t^2)/2$.
13. $a = -2\cos t \cot t$, $b = t + \cos t + \cos^3 t$.
14. $a = 144 \sin^3 t/13$, $b = 144 \cos^3 t/5$.
15. $a = 4(t + \sin t)$, $b = -4(1 - \cos t)$.
16. $a = \frac{1}{3}(2\cos t - \cos 2t)$, $b = \frac{1}{2}(2\sin t - \sin 2t)$.

EXERCISE XLVIII (Page 65)

1. $\partial u/\partial x = 3x^2 - 6xy + y^2$, $\partial u/\partial y = -3x^2 + 6xy + y^2$.
2. $\partial u/\partial x = 3x^2 - 3yz$, $\partial u/\partial y = 3y^2 - 3xz$, $\partial u/\partial z = 3z^2 - 3xy$.
5. $\partial u/\partial x = y/x$, $\partial u/\partial y = \log x$.
6. $\partial u/\partial x = -y\sin x + \cos y$, $\partial u/\partial y = \cos x - x\sin y$.
7. $\partial u/\partial x = e^y + ye^x$, $\partial u/\partial y = xe^y + e^x$.
14. $\partial z/\partial x = \tan x$, $\partial z/\partial y = -\tan y$.

EXERCISE XLIX (Page 66)

1. $-9x/4y$.
2. $(3x^2 - 2xy + y^2)/(x^2 - 2xy - 3y^2)$.
3. $(2 - 3x)/2y$.
4. $(ay - x^2)/(y^2 - ax)$.
5. $(y^2 - 1)/(3y^2 - 2xy + 1)$.
9. $-2\sin 2x/3\sin 2y$.
10. $(\cos y - y\cos x)/(\sin x + x\sin y)$.
12. $-(2 + 4ye^{xy}/3 + 4xe^{xy})$.
17. $(x + x\log x + y)/(x - x\log x)$.
19. $e^y/(2 - y)$.
23. y/x.
24. $(x\log a - y)/(x\log ax)$.

EXERCISE LI (Page 68)

1. $du = (2xy - 2y - 1)\,dx + (x^2 - 2x - 2y)\,dy$.
2. $du = [\cos(x + y) - \sin(x - y)]\,dx + [\cos(x + y) + \sin(x - y)]\,dy$.
3. $du = [y\cos x - \sin(x - y)]\,dx + [\sin x + \sin(x - y)]\,dy$.
4. $du = yze^{xyz}\,dx + xze^{xyz}\,dy + xye^{xyz}\,dz$.
10. $[2/(1 + x^2)]\,dx + [1/(1 + y^2)]\,dy$.
11. $(a^x b^y \log a + y\cos xy)\,dx + (a^x b^y \log b + x\cos xy)\,dy$.
12. $[1/(1 + x^2)]\,dx + [1/(1 + y^2)]\,dy$.
15. $du/dt = (2x - 2y)(dx/dt) + (2y - 2x)(dy/dt)$;
$du/dx = (2x - 2y) + (2y - 2x)(dy/dx)$.
16. $du/dt = (9x^2 - y^2 + 2xz)(dx/dt) + (2yz - 2xy)(dy/dt) + (x^2 + y^2)(dz/dt)$.
22. $du/dt = 2\tan t \sec^2 t - 2t$.
23. $du/dt = 2^{t\log t} \cdot \log 2\,(\log t + 1)$.

EXERCISE LII (Page 69)

1. (a) $-5/3$ units per sec.;
   (b) $5\sqrt{10}/3$ units per sec.,
   $dz/dx = -1/3$.
2. (a) 8; (b) $4\sqrt{5}$.
4. $dy/dt=2$, $dz/dt=24/7$; $v=4.44$.
5. $dx/dt = -1/4$, $dy/dt = -1/2$.
6. Vol., ± 2.16 cu. in.;
   surface, ± 1.44 sq. in.
7. (a) S, $.24\,\pi$ sq. in.;
   V, $.36\,\pi$ cu. in.;
   (b) S, 2/3%; V, 1%.
11. (a) .505 in.; (b) 101/132%.
12. $+1/20$.
14. (a) .3; (b) .5; (c) $6\frac{1}{4}$%.
15. (a) .0144;
    (b) 23/3600 = 23/36%.
16. .0522 ohms, 47/66%.
17. .0018.
18. .534.
19. .0204 sec., 37/64%.
21. (a) .908 in./sec.;
    (b) 8.88 in.^{2}/sec.
23. (a) $v = a\omega[\sin\theta + (c + a\sin\theta)\cos\theta/\sqrt{b^2 - (c + a\sin\theta)^2}]$, where θ is the angle through which the crank has turned;
    (b) $v = ca\omega/\sqrt{b^2 - c^2}$ and $a\omega$; (c) $a\omega/\sqrt{b^2 - c^2}$ and 0.
24. Volume, 3.0159 cu. in.; surface, 2.818 sq. in.
27. $144\,\pi/5$ cu. in.; (b) 3/10 or 30%.
29. (a) .0628%; (b) .1776%, 5.327 ft.
30. (a) $dJ = c/\cos^2\alpha \cdot d\alpha$; (b) $\alpha = 0°$ or $180°$; (c) $\alpha = 45°$.
31. (a) $dI_2 = 2\,a\,a_2 I_1/(a - a_2)^3 \cdot da_2$;
    (b) and (c) screen midway between the lights;
    (d) ± 1.4568 candle power; (f) ± 3.6134 candle power, 2.62%.

EXERCISE LIII (Page 74)

Of the first fifteen, all are exact except examples 6, 11 and 15.

EXERCISE LIV (Page 76)

The divergent series are examples 2, 7, 9, 14, 15, 20.

EXERCISE LV (Page 78)

1. $-1 < x < 1$.
2. $-1 \leqq x < 1$.
3. $-1 \leqq x \leqq 1$.
4. All values.
5. $-3 \leqq x < 3$.
6. $-1 < x < 1$.
9. $-\frac{1}{2} \leqq x \leqq \frac{1}{2}$.
14. $-2 \leqq x < 2$.
15. $-6 < x < 6$.
16. $-1/2 \leqq x \leqq 1/2$.
17. $-1 \leqq x < 1$.
18. $-2 \leqq x < 2$.

19–25. The student should observe that in all these the series converge for all values of $|x|$ less than a/b.

EXERCISE LVI (PAGE 79)

1. (a) $1 + 2(x - \pi/4) + 2(x - \pi/4)^2 + \frac{8}{3}(x - \pi/4)^3 + \cdots$;
   (b) $-1 + 2(x + \pi/4) - 2(x + \pi/4)^2 + \frac{8}{3}(x + \pi/4)^3 + \cdots$;
   (c) $\tan(-2) + \sec^2(-2)(x + 2) - \cdots$.
2. (a) $\frac{1}{2}\sqrt{3} + \frac{1}{2}(x - \pi/3) - \sqrt{3}/4 \cdot (x - \pi/3)^2 + \cdots$;
   (b) $-\frac{1}{2} + \sqrt{3}/2 \cdot (x + \pi/6) + \frac{1}{4}(x + \pi/6)^2 + \cdots$;
   (c) $1/\sqrt{2}(1 + x - x^2/2 - x^3/3! + \cdots)$.
3. $-33 - 13(x - 2) + 5(x - 2)^2 + 5(x - 2)^3 + (x - 2)^4$.
5. $-3 + 3(x + 1) - 4(x + 1)^2 + 8(x + 1)^3 - 5(x + 1)^4 + (x + 1)^5$.
6. $e^{-3}[1 + (x + 3) + \frac{1}{2}(x + 3)^2 + \cdots]$.
8. $\log 2 + (x - 1)/2 - (x - 1)^2/8 + (x - 1)^3/24 + \cdots$.
9. $\tan^{-1}3 + (x - 3)/10 - 3(x - 3)^2/100 + \cdots$.
14. $1/2 + (x + 1)/4 + (x + 1)^2/8 + (x + 1)^3/16 + \cdots$.
16. $4 - 3(x - 3) + 3(x - 3)^2 - 3(x - 3)^3 + \cdots$.

EXERCISE LVII (PAGE 80)

1. $2x - 8x^3/3! + 32x^5/5! + \cdots$.
2. $1 + 3x + 9x^2/2! + 27x^3/3! + \cdots$.
3. $\sqrt{3}/2 - x/4 - \sqrt{3}x^2/16 + x^3/96 + \sqrt{3}x^4/768 + \cdots$.
4. $-\sqrt{3}/2 + x/2 + \sqrt{3}x^2/4 - x^3/12 + \cdots$.
6. $x - x^2 + x^3/2! - x^4/3! + \cdots$.
9. $-x^2/2 - x^4/12 - x^6/45 + \cdots$.
10. $x - x^3/3 + x^5/5 - x^7/7 + \cdots$.
12. $x + x^2/2 - x^3/6 + x^4/12 + \cdots$.
16. $1 + x^2/2! + x^4/4! + x^6/6! + \cdots$.
18. $x - x^3/3! + 9x^5/5! + \cdots$.
23. $1 + x^2/2 - x^3/3 + x^4/8 + \cdots$.

EXERCISE LVIII (PAGE 81)

1, 2. .71934. 3. .54464.

4. $\log 2 = .69314805$, $\log 3 = 1.09861316$, $\log 5 = 1.60944011$, $\log 7 = 1.94591189$, $\log 11 = 2.38959929$, $\log 13 = 2.56495198$.

5. .02618. 6. 1.4844. 7. 1.34985. 10. .600889. 12. 2.006221. 16. 2.0409144. 17. 2.059756. 19. .00055 in. 30. -10.639. 33. 1.03543. 34. .97934.

EXERCISE LIX (PAGE 84)

1. 1. 2. 1. 3. 0. 4. 1/6. 5. 1/3. 6. $-1/3$. 7. 1/2. 8. 0. 9. 2. 10. 1/2. 11. ∞. 12. 2. 13. $1/\pi$. 14. ∞. 15. 0. 16. 1/3. 17. -1. 18. 1/2.

EXERCISE LX (PAGE 85)

1. $x^3 + xy^2 + 3x^2 + 4xy + y^2 + 7x + 4y + 5$.
2. $x - x^3/6 + x^2y + x^3y^2/2 - x^4y/6 + x^5/5! + \cdots$.
4. $y + xy\log a - y^2/2 + \frac{1}{2}x^2y(\log a)^2 - \frac{1}{2}xy^2\log a + \frac{1}{3}y^3 + \cdots$.
6. $(1/2!)(2xy) + (1/4!)(x^4 + 6x^2y^2 + y^4) - (1/6!)(x^6 + 15x^4y^2 + 15x^2y^4 + y^6) + \cdots$.
8. (a) $(x + y) - (1/3!)(x^3 + 3x^2y + 3xy^2 + y^3) + \cdots$;
   (b) $\sin(x+y) + (h+k)\cos(x+y) - (1/2!)(h^2 + 2hk + k^2)\sin(x+y)\cdots$.
9. $\cos(xy) - (hy + kx)\sin(xy) + \cdots$.
10. $e^x \sin y(1 + h + h^2/2! + \cdots)(1 - k^2/2! + k^4/4!\cdots) + e^x\cos y(1 + h + h^2/2! + \cdots)(k - k^3/3! + k^5/5!\cdots) + \cdots$.
11. $\log(e^x + e^y) + (he^x + ke^y)/(e^x + e^y) + e^xe^y(h - k)^2/2(e^x + e^y)^2 + \cdots$.

EXERCISE LXI (PAGE 86)

1. Min., $x = 8/3$, $y = 2/3$.
2. Min., $x = \frac{27}{2}$, $y = 5$.
3. Min., $x = y = \frac{1}{2}\sqrt[3]{2}$.
4. Min., $x = -1$, $y = 3$.
5. Max., $x = y = \pi/3$.
6. Max., $x = 1$, $y = 2$.
7. Max., $x = y = \pi/3$.
8. Max., $x = y = 2$.
16. $4 \times 4 \times 4$ ft.
17. Equal parts.

EXERCISE LXII (PAGE 88)

1. $x^2 + 4y = 0$.
2. $27x^2 = 4y^3$.
3. $27y = x^3$.
4. $xy^2 - x^2 - 2x - 1 = 0$.
5. $y^2 + 4x = 0$.
8. $y^2 - 4x - 4 = 0$.
10. $x^2 - 2xy + y^2 = 8$.
12. $x^4 + 4y = 0$.
13. $4x^3 = 27y$.
14. $y = \pm 2x$.
19. $y^2 = 16x$.
20. $x^2 + 4y^2 = 64$.
21. $x^2 + 4y^2 = 4$.
22. $25x^2/81 + 25y^2/256 = 1$.
27. $x^{2/3} + y^{2/3} = 4$.
28. $y^2 = 8x + 8$.

EXERCISE LXIII (PAGE 89)

(A) 1. $3x - 3y - 4 = 0$.
2. $x - y + 3 = 0$, $2x - 3y - 13 = 0$.
3. $x - 2y = 0$.
4. $x = 4$, $x - y + 4 = 0$, $x + y + 4 = 0$.
5. $x - y = 0$, $x = \pm 1$.
6. $x = 0$, $y = \pm 1$.
7. $x = 0$, $y = 0$, $x + y = 0$.
8. $x - y - 2 = 0$, $x + y = 0$.
9. $x - y - 4 = 0$, $x - y + 4 = 0$, $x + 4 = 0$.
10. $x - y - 2 = 0$, $x - y + 2 = 0$.
11. $x + y = 0$, $x - y = 0$.
12. $3x + 3y - 8 = 0$.
28. $4x - 4y + 3 = 0$, $84x + 28y + 9 = 0$, $7x - 14y + 1 = 0$.
29. $x - y = 0$, $x + y + 2 = 0$, $2x - y + 1 = 0$.
30. $x = 2$, $2x + 8y + 1 = 0$, $6x - 40y + 9 = 0$.
31. $x + 1 = 0$, $6x - 9y + 28 = 0$.

(B) 1. $3\rho\sin(\pi/3+\theta)+4\sqrt{3}=0$, $3\rho\sin(\pi/3-\theta)+4\sqrt{3}=0$.
2. $\sqrt{3}\rho\sin(\pi/6+\theta)=2$, $\sqrt{3}\rho\sin(5\pi/6+\theta)+2=0$.
5. $\theta=0$.
6. $\rho\cos\theta=8$.
9. $\rho\sin(\pi/4-\theta)=4$, $\rho\sin(5\pi/4-\theta)=4$.
13. $\rho\cos\theta=4$.
14. $\rho\cos\theta=-4$.
15. $\sqrt{2}\rho\sin(\theta-\pi/4)=1$.
16. $\rho\sin\theta=2\pi$, etc.
17. $\rho\cos\theta=-4$.
18. $\rho\sin\theta=2$.
19. $\rho\sin\theta=1$.
20. $\rho\sin\theta=\pi$.
30. $\rho\sin\theta=\pm 2$, $\rho\cos\theta=\pm 2$.

(C) 1. (0, 0) conjugate point.
2. As 1.
3. Cusp at (0, 0).
4. Double point at (0, 0).
7. Double point at (– 2, 3).
8. Conjugate point at (0, 2).
9. Cusp at (1, 1).
10. (0, 3), (– 3, 0), and (– 6, 3) are double points.
11. (0, 1), (0, – 1), (1, 0), and (– 1, 0) are all double points, hence the curve degenerates.
14. (2, 0) and (– 2, 0) are both double points.
15. (2, 0), (– 2, 0), and (0, – 2) are double points.
16. Cusp at (3, 2). 17. Conjugate point at (0, – 1). 18. Cusp at (0, 0).

EXERCISE LXIV (Page 93)

1. $(x+1)/1=(y-5)/4=(z-4)/4$; $x+4y+4z=40$.
2. $x/1=(y-2)/2=(z+1)/-1$; $x+2y-z=5$.
3. $(x-1)/2=(y-1)/3=(z-5)/2$; $2x+3y+2z=15$.
4. $(x-4)/4=(y-1)/-1=(z-4)/8$; $4x-y+8z=47$.
7. $(x-1)/2=(y-1)/-1$, $z=1$; $2x-y-1=0$.
8. $z-1=0$, $y-1=0$; $x=0$.
9. $x=2$, $y/-3=\pi(z-1)/2$; $3\pi y-2z+2=0$.
11. $(x-3)/2=(y-2)/-3$, $z+6=0$; $2x-3y=0$.
12. $(x-2)/5=(y-1)/-11=(z-4)/-2$; $5x-11y-2z+9=0$.
14. $(x-2)/16=(y-4)/-5=(z-2)/6$; $16x-5y+6z=24$.
15. $(x-2)/6=(y-1)/-21=(z-3)/1$; $6x-21y+z+6=0$.

EXERCISE LXV (Page 94)

1. $6x+2y+3z=49$; $(x-6)/6=(y-2)/2=(z-3)/3$.
2. $4x+2y-z=6$; $(x-2)/4=(y-1)/2=(z-4)/-1$.
7. $13x+15y+z+15=0$; $(x-2)/13=(y+3)/15=(z-4)/1$.
8. $2x+2y-z+8=0$; $(x-1)/2=(y+2)/2=(z-6)/-1$.
9. $4x+y+z-13=0$; $(x-2)/4=(y-1)/1=(z-4)/1$.
11. $\cos^{-1}19/3\sqrt{138}$.
12. $\cos^{-1}19/7\sqrt{29}$.
13. Each is $3x+4y+3z-20$, ∴ curve cuts surface orthogonally.
14. $3x+4y+6z=22$, $6x+y-z=11$; angle is $\cos^{-1}16/\sqrt{2318}$.
15. Tangent plane to each surface is $2x+3y+2z=9$.

EXERCISE LXVI (PAGE 96)

1. $\frac{1}{3}x^3$. **2.** $\frac{3}{8}x^{\frac{8}{3}}$. **3.** $\frac{3}{4}x^{\frac{4}{3}}$. **4.** $-1/4x^4$. **8.** $-1/2\sqrt{x}$. **9.** $2\sqrt{2}x^{3/2}/3$.

13. $x+\frac{2}{3}x^{3/2}$.
14. $\frac{2}{3}x^{\frac{3}{2}}+\frac{1}{3}x^3$.
15. $\frac{1}{2}x^2-2\sqrt{x}$.
16. $\frac{3}{4}x^{\frac{4}{3}}+\frac{3}{2}x^{\frac{2}{3}}$.
17. $-\frac{1}{4}(1-x)^4$.
18. $x+\frac{4}{3}x^{\frac{3}{2}}+\frac{1}{2}x^2$.
19. $x-3x^{2/3}+3x^{1/3}$.
20. $1+2/x-1/3x^3$.
25. $\frac{2}{3}x^{3/2}-\frac{2}{7}x^{7/2}$
26. $\frac{2}{3}x^{3/2}-\frac{4}{5}x^{5/2}+\frac{2}{7}x^{\frac{7}{2}}$.
29. $\frac{3}{2}z^{2/3}-\frac{12}{7}z^{7/6}+\frac{3}{5}z^{5/3}$.
34. $\frac{1}{12}(3n+2)^4$.
35. $\frac{1}{30}(3-5x)^{-6}$.
36. $-7/4(1+2x)^2$.
39. $\frac{2}{3}\sqrt{1+3\theta}$.
41. $\frac{1}{16}(s^2+1)^8$.
43. $-1/6(3+2x^3)$.
45. $-\frac{5}{6}(1-2x^2)^{3/2}$.
47. $\frac{1}{5}(2x^{3/2}+1)^{5/3}$.
49. $\frac{9}{20}(2x^{5/3}+5)^{2/3}$.
51. $\sqrt{x^2+2x}$.
55. $-\frac{4}{3}(1-\sqrt{x})^{3/2}$.
56. $-4\sqrt{1-\sqrt{x}}$.
57. $\frac{1}{9}(1+e^{3x})^3$.
58. $-\frac{2}{3}(1-e^x)^{3/2}$.
59. $\frac{1}{2}(2+\log x)^2$.
61. $\frac{2}{3}\sin^3\frac{1}{2}x$.
62. $\frac{1}{4}\tan^2 2x$.
63. $-\frac{2}{3}\sqrt{\cos 3x+5}$.
64. $-1/(1+\tan x)$.

EXERCISE LXVII (PAGE 98)

1. $\log(x+1)$. **2.** $\frac{1}{3}\log(1+3x)=\log\sqrt[3]{1+3x}$. **3.** $-\log\sqrt{1-2x}$.
4. $-\frac{1}{14}\log(2-7x^2)$.
5. $\frac{2}{3}\log(x^3+1)$.
6. $-\frac{7}{20}\log(2-5x^4)$.
7. $\log(x^{\frac{5}{2}}+2)^2$.
10. $\frac{2}{3}\log(1+x^{3/2})$.
11. $\log(2+\sqrt{x})^2$.
16. $\log\sqrt{2+e^{2x}}$.
17. $-\log(3+e^{-x})$.
18. $-\log(1-3^{3x})/3\log 3$.
21. $-\frac{1}{2}\log(1-\log^2 x)$.
23. $\log(1+\sin^2 x)$.
24. $-\frac{3}{10}\log(1-2\sin 5x)$.
29. $\frac{1}{2}\log(\log\tan x)$. **30.** $\log(\text{arc}\tan x)$.

EXERCISE LXVIII (PAGE 99)

1. $2^x/\log 2$.
2. $5^{2x}/2\log 5$.
3. $-1/3^x\log 3$.
4. $-e^{-x}$.
7. $\frac{1}{5}e^{(5x+1)}$.
9. $\frac{1}{2}e^{2x-1}$.
10. $\frac{1}{2}e^{x^2}$.
12. $-e^{1/x}$.
13. $\frac{1}{2}e^{\sin 2x}$.
14. $1/e^{\cos x}$.
15. $\frac{1}{2}e^{2\tan x}$.
18. $-\frac{1}{4}e^{\log(1-4x)}$.
21. $\frac{1}{2}e^{2x}+\frac{1}{2}e^{4x}+\frac{1}{6}e^{6x}$.
22. $\frac{1}{4}e^{4x}+2x-\frac{1}{4}e^{-4x}$.
26. $\frac{1}{2}e^{2x}-2e^{x+\frac{1}{2}}+ex$.
29. $-\frac{1}{2}e^{2/x}-2x^{-\frac{1}{2}}$.
31. $e^x-\log(e^x+1)$.
32. $-\frac{1}{2}e^{2x}-e^x-\log(1-e^x)$.

EXERCISE LXIX (PAGE 100)

1. $-\frac{1}{2}\cos 2x$.
2. $\frac{1}{3}\log\sec 3x$.
3. $\frac{1}{5}\log(\sec 5x+\tan 5x)$.
4. $3\sin(x/3)$.
7. $\sqrt{2}\log\sec(x/\sqrt{2})$.
10. $\frac{1}{2}\sin(1+2x)$.
11. $-\log\sin(1-x)$.
13. $\frac{1}{5}\tan 5x$.
14. $\frac{1}{2}\sec 2x$.
15. $2\log(\csc x/2-\cot x/2)$.
16. $\frac{1}{2}\sin x^2$.
18. $2\sin\sqrt{x}$.
19. $\frac{1}{3}\tan x^3$.
20. $\cot 1/x$.
22. $\frac{1}{3}\sin 3x$.
25. $2\tan x/2-x$.
28. $-\frac{1}{3}\cos 3x$.
29. $\frac{1}{2}\tan 2x$.
30. $\frac{1}{5}\sec 5x$.
34. $\sin x$.
35. $2x-\tan x$.
37. $\log\sec e^x$.
38. $\cos e^{-x}$.
42. $x+\log(\sec 2x+\tan 2x)+\frac{1}{2}\tan 2x$.
44. $x-\frac{1}{4}\cos 4x$.
50. $\frac{1}{2}\tan x$.
51. $\frac{1}{3}(\tan 3x+\sec 3x)$.
52. $x-\frac{1}{3}(\csc 3x-\cot 3x)$.
53. $\frac{1}{2}(\sec 2x+\tan 2x)-x$.

EXERCISE LXX (Page 102)

1. $\arctan x$.
2. $\frac{1}{2}\log[(x-1)/(x+1)]$
3. $\frac{1}{2}\log[(x+1)/(x-1)]$.
5. $\frac{1}{6}\log[(3x-1)/(3x+1)]$.
6. $\frac{1}{12}\log[(3x+2)/(3x-2)]$.
7. $1/2\sqrt{5}\cdot\log\left[(x-\sqrt{5})/(x+\sqrt{5})\right]$.
8. $1/\sqrt{2}\cdot\arctan x\sqrt{2}$.
11. $1/3\sqrt{7}\cdot\arctan\left(\sqrt{7}\theta/3\right)$.
13. $\frac{1}{2}\log[x/(x+2)]$.
14. $\frac{1}{4}\log[x/(x-4)]$.
17. $\frac{1}{2}\log[(x+1)/(x+3)]$.
18. $-\frac{1}{3}\tan^{-1}[(x-1)/3]$.
19. $\arctan[(x-4)/3]$.
22. $1/\sqrt{3}\cdot\arctan(x-3)/\sqrt{3}$.
24. $2/\sqrt{3}\cdot\arctan(2x+3)/\sqrt{3}$.
25. $\frac{1}{4}\arctan(2x+1)/2$.
26. $\frac{1}{6}\log[(3x-4)/3x+2]$.
30. $\sqrt{2}/2\cdot\arctan(4t+1)/2\sqrt{2}$.
37. $\frac{1}{2}\arctan(x^2+1)$.
38. $\frac{1}{12}\log[(x^3-5)/(x^3-1)]$.
39. $\log(\sqrt{x}-1)/(\sqrt{x}+1)$.
40. $3\arctan x^{\frac{1}{3}}$.
41. $\frac{1}{2}\arctan e^x$.
43. $\arctan(\log x)$.
45. $\frac{1}{9}\arctan(\sin 3x)$.
48. x.

EXERCISE LXXI (Page 103)

3. $\frac{1}{3}\arcsin 3x/2$.
4. $\frac{1}{2}\log\left(2x+\sqrt{4x^2+3}\right)$.
5. $1/\sqrt{7}\cdot\arcsin\sqrt{7}x/3$.
6. $2/\sqrt{5}\cdot\log\left(\sqrt{5}x+\sqrt{5x^2+2}\right)$.
7. $\arcsin(x-1)$.
9. $\frac{1}{2}\log\left(x+\frac{7}{8}+\sqrt{x^2+7x/4}\right)$.
11. $\arcsin(x-1)/4$.
14. $\log\left(x-3+\sqrt{11-6x+x^2}\right)$.
17. $2/\sqrt{3}\cdot\arcsin(3x+2)/\sqrt{19}$.
18. $6/\sqrt{7}\cdot\arcsin(7x+4)/\sqrt{79}$.
19. $\frac{1}{2}\log\left(e^{2x}+\sqrt{e^{4x}-4}\right)$.
21. $\arcsin\left(\log x/\sqrt{3}\right)$.
22. $2\log\left(\sin x/2+\sqrt{\sin^2(x/2)+5}\right)$.

EXERCISE LXXII (Page 104)

3. $\frac{1}{4}\log(4x^2+1)+\frac{3}{2}\arctan 2x$.
4. $\frac{1}{6}\log[(3x-1)/(3x+1)]-\frac{1}{3}\log(9x^2-1)$.
5. $\frac{1}{3}\log(3x^2-2)-\sqrt{6}/4\cdot\log\left[(\sqrt{3}x-\sqrt{2})/(\sqrt{3}x+\sqrt{2})\right]$.
9. $\frac{1}{2}\log(x^2+2x)+\frac{1}{2}\log[x/(x+2)]$.
10. $-\frac{1}{2}\log(6x-x^2)-\log(x-6)/x$.
13. $\log(x^2+2x+5)+\frac{3}{2}\arctan(x+1)/2$.
18. $-\frac{1}{2}\log(2-6x-x^2)+3/2\sqrt{11}\cdot\log\left[(x+3-\sqrt{11})/(x+3+\sqrt{11})\right]$.
21. $-\frac{1}{8}\log(4x^2-4x-3)+\frac{1}{16}\log[(2x-3)/(2x+1)]$.
23. $-\frac{1}{6}\log(1-6x-9x^2)+1/2\sqrt{2}\cdot\log\left[(3x+1-\sqrt{2})/(3x+1+\sqrt{2})\right]$.
27. $\frac{1}{6}\log(3x^2+2x-1)+\frac{5}{12}\log[(3x-1)/(x+1)]$.
29. $-\frac{1}{6}\log(3x^2-4x+3)+\sqrt{5}/15\cdot\arctan(3x-2)/\sqrt{5}$.
34. $-\frac{1}{2}\log(4-5x-3x^2)+\dfrac{19}{2\sqrt{73}}\log\left[(6x+5-\sqrt{73})/(6x+5+\sqrt{73})\right]$.

EXERCISE LXXIII (Page 106)

1. $2\sqrt{x^2-1}+\log(x+\sqrt{x^2-1})$.
2. $-\sqrt{1-x^2}-\text{arc}\sin x$.
3. $\frac{2}{3}\sqrt{9x^2+1}+\frac{5}{3}\log(3x+\sqrt{9x^2+1})$.
4. $\frac{1}{2}(\text{arc}\sin 2x+\sqrt{1-4x^2})$.
7. $\sqrt{x^2+2x}+2\log(x+1+\sqrt{x^2+2x})$.
8. $-\sqrt{4x-x^2}+4\,\text{arc}\sin(x-2)/2$.
11. $-\sqrt{27+6x-x^2}+3\,\text{arc}\sin(x-3)/6$.
14. $3\sqrt{19-5x+x^2}+\frac{19}{2}\log(x-5/2+\sqrt{19-5x+x^2})$.
17. $\frac{3}{4}\sqrt{4x^2-4x+5}-\frac{1}{4}\log(2x-1+\sqrt{4x^2-4x+5})$.
20. $-2\sqrt{12x-4x^2-5}+\frac{9}{2}\,\text{arc}\sin(2x-3)/2$.
21. $\frac{1}{3}\sqrt{3x^2+6x-2}-\sqrt{3}\log(x+1+\sqrt{x^2+2x-\frac{2}{3}})$.

EXERCISE LXXIV (Page 107)

1. $\frac{1}{2}\sin^2 2x$.
2. $\frac{1}{9}\sin^3 3x$.
3. $-\cos^3 x/3$.
5. $\cos^3 x/3-3\cos x/3$.
6. $\frac{3}{2}\sin 2x/3-\frac{1}{2}\sin^3 2x/3$.
7. $\frac{1}{4}\sec^2 2x$.
9. $\frac{2}{5}\cos^5 x/2-\frac{2}{3}\cos^3 x/2$.
11. $2/7\pi\cdot\cos^{\frac{7}{2}}\pi x-2/3\pi\cdot\cos^{\frac{3}{2}}\pi x$.
13. $-\frac{2}{3}\sqrt{\cos 3x}$.
15. $\frac{10}{3}\cos^3 x/5-5\cos x/5-\cos^5 x/5$.
17. $-2\sqrt{\cos x}$.
19. $\frac{1}{2}\cos^2 x-\log\cos x$.
21. $x/2-\frac{3}{4}\sin 2x/3$.
23. $x/8-\frac{1}{72}\sin 8x$.
25. $3x/8-\frac{1}{2}\sin x+\frac{1}{16}\sin 2x$.
27. $\frac{1}{16}\sin^2 x-\frac{1}{8}\cos x$.
29. $\frac{5}{16}x-\frac{1}{4}\sin(2-2x)-\frac{3}{64}\sin(4-4x)+\frac{1}{24}\sin^3(2-2x)$.
33. $3x/2-\cos 2x-\frac{1}{8}\sin 4x$.
35. $5x/8-\frac{2}{3}\cos^3 x+\frac{1}{32}\sin 4x$.
37. $\frac{1}{4}\sin^4 x$.
39. $\frac{2}{3}\sin^3 x$.

EXERCISE LXXV (Page 109)

1. $\frac{1}{3}\tan^3 x$.
2. $\frac{2}{3}\tan^{\frac{3}{2}} x$.
3. $\frac{1}{4}\tan^2 2x$.
4. $\frac{1}{6}\tan^2 3x-\frac{1}{3}\log\sec 3x$.
5. $\frac{1}{9}\sec^3 3x-\frac{1}{3}\sec 3x$.
6. $3\sec^3 x/3$.
7. $\frac{1}{12}\tan^3 4x-\frac{1}{4}\tan 4x+x$.
9. $\frac{2}{3}\tan^{3/2}x$.
11. $-\frac{1}{6}\cot^2 3x$.
15. $-\frac{5}{4}\cot^4 x/5+\frac{5}{2}\cot^2 x/5-5\log\csc x/5$.
21. $\frac{1}{7}\tan^7 x+\frac{1}{5}\tan^5 x$.
23. $\tan x-2\cot x-\frac{1}{3}\cot^2 x$.
24. $-\cot x+2\tan x+\frac{1}{3}\tan^3 x$.

EXERCISE LXXVI (Page 111)

2. $\frac{1}{2}\,\text{arc}\sec x/2$.
3. $\sqrt{x^2-3}/3x$.
4. $-\sqrt{5-x^2}/5x$.
5. $x/2\sqrt{x^2+2}$.
7. $-\sqrt{4-x^2}/x^2-\text{arc}\sin x/2$.
8. $\sqrt{x^2-16}-4\,\text{arc}\sec x/4$.
10. $\frac{1}{3}\log(x/(3+\sqrt{9-x^2}))$.
12. $\log(x+\sqrt{x^2-36})-\sqrt{x^2-36}/x$.
13. $(x^2-8)^{3/2}/24x^3$.
15. $\sqrt{x^2-3}(2x^2+3)/27x^3$.
16. $-(2-x^2)^{3/2}(3x^2+4)/15$.
22. $\frac{1}{16}\,\text{arc}\sec x/2+\sqrt{x^2-4}/8x^2$.
23. $(x^2-2)\sqrt{x^2+1}/3$.
24. $(x^2+2)\sqrt{x^2-1}/3$.

EXERCISE LXXVII (Page 113)

1. $y = x^2 + 4$.
2. $2x = \log y + 1$.
3. $x^2 = \log y^2 + 4$, or $y = e^{\frac{1}{2}(x^2-4)}$.
5. $x^2 - y^2 - 4x + 4 = 0$.
7. $x^2 + y^2 - 4x + 6y - 12 = 0$.
8. $y = 1/(1 - \log x)$.
9. $16y = (x^2 + 4)^2$.
12. $y(1 + \arcsin x) + 1 = 0$.
13. $x^2 - y^2 + 1 = 0$.
17. $2\tan y = x^2 - 4$.
19. $y(1 + 2\cos x) - 2 = 0$.
21. $x \tan y = x - 1$.
23. $y = \frac{1}{2}\sin(4\sqrt{x} + \pi/6)$.
26. $15y^2 - 2x^2y^2 - 16y + 4 = 0$.
28. $v = F/k \cdot (1 - e^{kt/M})$.
29. $v = v_0 e^{-kt/M}$.
30. $v_{max} = \sqrt{F/k}$.
31. 18.97 ft./sec., 18.78 ft./sec.
36. $y = \pi^2(x^2 + z^2)/4$.
37. $T = 50e^{s/8}$.

EXERCISE LXXVIII (Page 115)

1. $x + \log y = C$.
2. $xy + Cx - 1 = 0$.
3. $x^2 - y^2 = C$.
9. $y = \log x - \frac{1}{2}x^2 + C$.
11. $3y = (1 + x)^3 + C$.
13. $y + x/2 + \frac{1}{4}\sin 2x + C = 0$.
15. $x^{3/2} - y^{3/2} = C$.
17. $x - y + Cxy = 0$.

EXERCISE LXXIX (Page 115)

1. 1/21.
3. -8.
5. 2.
7. 1.
9. 24.
10. 9.
11. $\sqrt{8}(\sqrt{3} - 1)$.
13. 7/5.
15. $\log\sqrt{3}$.
17. $\log\sqrt{2}$.
19. $\pi/18$.
20. $2\pi/3\sqrt{3}$.
25. $\pi/4$.
27. $2\pi/3\sqrt{3}$.
31. $-\pi/2$.
33. 1/3.
35. 3/32.
37. $\pi/12$.
39. 4/3.
40. $2\pi a^2$.
41. No finite limit.
42. 12.

EXERCISE LXXX (Page 117)

1. $10\frac{2}{3}$.
2. $1\frac{1}{3}$.
3. 8.
4. $21\frac{1}{3}$.
5. 1.
7. $10\frac{2}{3}$.
9. 3.
11. log 16.
13. 6.524.
15. $26\frac{2}{3}$ acres.
16. $12,800.

EXERCISE LXXXI (Page 119)

1. $x + \log(x - 1)$.
3. $-2x - \log\sqrt{3 - 2x}$.
5. $x^2/2 + x + \log(x - 1)^2$.
9. $x^2 - \log\sqrt{x^2 + 1} + \arctan x$.
11. $x^2/2 - x + 2/\sqrt{3} \cdot \arctan[(2x + 1)/\sqrt{3}]$.
12. $x + 2\log(x^2 - 2x + 1) + 4/(x - 1)$.
13. $\log x^2(1 + x)^3/(1 - x)^5$
15. $\log(2x - 1)^3(2x + 1)/x$.
19. $\log(3x + 2)^{2/3}(2x + 3)^{3/2}/x$.
23. $\log[(x + 1)^{1/2}(x - 3)^{1/4}/(x^2 + 2x - 3)^{3/8}]$.
29. $-3/x + 4\log(x + 1)/x$.
30. $x - 2/x + \log x^2/(x - 1)$.

33. $\log(x^2+x)-1/(x+1)$. **41.** $\log(x-1)^2-(x-2)/(x^2-x)$.
35. $\log[(x-1)^{3/2}/(x-3)^{\frac{1}{2}}]-13/(x-3)$. **47.** $\log(x/\sqrt{x^2+1})$.
37. $\log(x+1)-(4x+7)/2(x+1)^2$. **49.** $\log x+2\arctan x$.
51. $\log(2x^2+3)/x^3+1/\sqrt{6}\cdot\arctan\sqrt{2}x/\sqrt{3}$.
52. $\log[(x-2)/\sqrt{x^2+1}]+\arctan x$. **62.** $2\arctan x/2-\arctan x$.
64. $1/\sqrt{3}\cdot\arctan x/\sqrt{3}+\frac{1}{6}\log(3x^2+1)$.
66. $\frac{1}{2}\log(x^2+2)(x^2+1)+\sqrt{2}\arctan x\sqrt{2}-\arctan x$.
72. $\frac{7}{2}+\log 4$. **74.** $\log 5$.
73. $\pi/6+\log\sqrt{(1+\sqrt{3})/(1-\sqrt{3})}$. **75.** $\frac{3}{4}+4\log\frac{5}{6}$.

EXERCISE LXXXII (Page 122)

1. $2\sqrt{x}-2\log(1+\sqrt{x})$. **2.** $-\sqrt{x}-\frac{1}{2}\log(1-2\sqrt{x})$. **3.** $2\arctan\sqrt{x}$.
5. $\log(x+2\sqrt{x}+5)-\arctan\frac{1}{2}(\sqrt{x}+1)$.
7. $2\log(\sqrt{x}-4)$. **16.** $\frac{2}{3}(1+x)^{3/2}-2(1+x)^{1/2}$.
9. $3\log(\sqrt[3]{x}/(1-\sqrt[3]{x}))$. **17.** $\log 3/2$.
13. $6-18\tan^{-1}1/3$. **18.** $\log[(\sqrt{1-x}-1)/(\sqrt{1-x}+1)]$.
15. $116/15$. **19.** $\pi/9$.
25. $\frac{2}{3}\sqrt{2x+3}+\sqrt{15}/9\cdot\log[(\sqrt{6x-9}-\sqrt{5})/(\sqrt{6x-9}+\sqrt{5})]$.
32. $2/\sqrt{1-x}+2\log[(\sqrt{1-x}-3)/(\sqrt{1-x}+3)]$.
33. $-2(x+1)-2\arctan\sqrt{x+1}$; $\frac{4}{3}-\pi/6$.
36. $\frac{1}{2}\log(\sqrt{x+1}-1)/(\sqrt{x+1}+1)-1/\sqrt{3}\cdot\arctan\sqrt{(x+1)/3}$.
45. $-\frac{1}{2}(\sqrt{4+x^4}/x^2)-\frac{1}{4}\log[(\sqrt{4+x^4}/x^2-1)/(\sqrt{4+x^4}/x^2+1)]$.

EXERCISE LXXXIII (Page 124)

1. $\log(1+\tan x/2)$. **6.** $\frac{2}{3}\log[(\tan x/2+3)/(\tan x/2-3)]$.
3. $\frac{1}{2}\log\tan(x/2)-\frac{1}{4}\tan^2(x/2)$. **9.** $1/\sqrt{2}\cdot\arctan(1/\sqrt{2}\cdot\tan\frac{1}{2}x)$.
5. $\frac{2}{3}\arctan(\frac{1}{3}\tan x/2)$. **13.** $\arctan(2\tan x/2+1)$.
15. $-x/3+\frac{5}{6}\arctan(2\tan x/2+1)$.
17. $\arctan(\tan^2 x/2)$.
20. $\frac{2}{5}\arctan(\tan x/2)+\frac{4}{15}\log[(\tan x/2-3)/(\tan x/2+3)]$.

EXERCISE LXXXIV (Page 125)

1. $(4-x^2)^{3/2}/12x^3$. **9.** $-\arcsin[(1+x)/2x]$.
2. $-\sqrt{x^2+5}/5x$. **10.** $\log[(1+2x+\sqrt{1+4x+5x^2})/x]$.
3. $-\frac{1}{2}\arcsin 2/3x$. **13.** $-\frac{1}{2}\arcsin[(2-x)/x\sqrt{2}]$.
5. $-\sqrt{1+2/x}$. **14.** $-\frac{1}{3}\arcsin[(3+x)/x\sqrt{3}]$.
7. $\frac{2}{9}(1-3/x)^{3/2}$. **15.** $-\frac{1}{4}\log[(4+3x+\sqrt{16+24x+3x^2})/x]$.
17. $-\sqrt{1+2x+3x^2}/x+\log[(1+x+\sqrt{1+2x+3x^2})/x]$.
19. $\sqrt{27x^2+6x-1}/x-3\arcsin[(1-3x)/6x]$.
21. $-\frac{2}{3}\arctan\frac{1}{3}\sqrt{(1+x)/3x}$. **25.** $-\frac{2}{3}\sqrt{x/(x+3)}$.
22. $\log[(\sqrt{x+1}-\sqrt{x})/(\sqrt{x+1}+\sqrt{x})]$. **26.** $\frac{1}{5}[(x-2)/x]^{5/2}$.

EXERCISE LXXXV (PAGE 126)

1. $2 + \log 3/2$. **2.** $2 \log 4/3$. **3.** $46\frac{2}{5}$. **5.** $10\frac{2}{3}$. **10.** 3.
11. $1/\sqrt{2} \cdot \log(5 + 2\sqrt{7})/(7 + 2\sqrt{13}) = -.226$.
13. $-7\pi/6$. **16.** $\frac{1}{2}\sqrt{3} + \pi/3$. **19.** $\frac{17}{3}\sqrt{5}$. **22.** 18.
14. $-4\pi/3$. **17.** $2/3$. **20.** 4π. **23.** $\pi a^2/4$.
15. $\pi/12 - \sqrt{3}/8$. **18.** $\sqrt{3} - \pi/3$. **21.** $2\sqrt{3} - \pi/3$. **24.** $5/27$.

EXERCISE LXXXVI (PAGE 127)

1. $2(\sin\sqrt{x} - \sqrt{x}\cos\sqrt{x})$. **3.** $\text{arc}\sin x + \sqrt{1 - x^2}$. **5.** $x - \log(e^x + 1)$.
2. $\sin x - \cos x$. **4.** $\frac{2}{3}\,\text{arc}\sin x^{3/2}$. **6.** $\text{arc}\tan e^x$.
7. $2/3[(x + 3)^{3/2} + (x + 2)^{3/2}]$.
8. $-\frac{1}{2}[x^2 + x\sqrt{x^2 + 1} + \log(x + \sqrt{x^2 + 1})]$.
9. $\log[(\sqrt{e^x + 1} - 1)/(\sqrt{e^x + 1} + 1)]$. **11.** $\frac{2}{3}(1 + \log x)^{3/2}$.
14. $\frac{1}{2}(x + 1)\sqrt{x^2 + 2x + 2} + \frac{1}{2}\log(x + 1 + \sqrt{x^2 + 2x + 2})$.
15. $\frac{1}{3}(x^2 - 2)\sqrt{1 + x^2}$.
16. $\log(\sqrt{1+x} - \sqrt{1-x})/(\sqrt{1+x} + \sqrt{1-x}) + 2\,\text{arc}\tan\sqrt{(1-x)/(1+x)}$.
21. $(x^3 + 4x^2 + 24x + 180)\sqrt{x^2 - 4x - 5} + 450\log(x - 2 + \sqrt{x^2 - 4x - 5})$.
22. $(6 - 2x)\sqrt{5 - 4x - x^2} + 25\,\text{arc}\sin\frac{1}{3}(x + 2)$.
23. $2/\sqrt{11}\cdot\text{arc}\tan[(4\tan x - 3)/\sqrt{11}(4 + 3\tan x)]$.
24. $\frac{1}{24}(4x^3 - 6x^2 - 6x - 1)/(1 + 4x)^{3/2}$. **27.** $r(\pi - 2)/2$.
30. $\frac{1}{16}[3a^2\,\text{arc}\sin(x^{1/3}/a^{1/3}) - (3a^{4/3}x^{1/3} - 6a^{2/3}x)\sqrt{a^{2/3} - x^{2/3}}$
$+ 8x(a^{2/3} - x^{2/3})^{\frac{3}{2}}]$.

EXERCISE LXXXVII (PAGE 129)

1. $-\frac{1}{2}x\cos 2x + \frac{1}{4}\sin 2x$. **5.** $\frac{1}{4}x^2 - \frac{1}{2}x\sin x - \frac{1}{2}\cos x$.
2. $\frac{1}{3}x\sin 3x + \frac{1}{9}\cos 3x$. **7.** $-\frac{1}{2}x^2\cos 2x + \frac{1}{2}x\sin 2x + \frac{1}{4}\cos 2x$.
3. $\frac{1}{2}x\tan 2x + \frac{1}{4}\log\cos 2x$. **9.** $\frac{1}{8}(3\sin x\sin 3x + \cos x\cos 3x)$.
13. $x\,\text{arc}\tan x/3 - \frac{3}{2}\log(x^2 + 9)$.
17. $\frac{1}{2}x^2\,\text{arc}\tan(1 - x) + x/2 + \log(x^2 - 2x - 2)$.
19. $\frac{1}{4}(x^4 - 1)\,\text{arc}\tan x + \frac{1}{12}(3x - x^3)$.
21. $\log(x/\sqrt{1 + x^2}) - \text{arc}\tan x/x$.
25. $(x - 1)\log(1 - \sqrt{x}) - \frac{1}{2}(x + 2\sqrt{x})$.
29. $x + \cos x[1 - \log(1 + \sin x)]$.
31. $\frac{1}{4}(3 - 2x)e^{2x+1}$. **42.** $\frac{1}{9}e^{-x}(-\sin 3x - 3\cos 3x)$.
33. $\frac{1}{3}e^{3x}(x^2 - \frac{2}{3}x + \frac{2}{9})$. **45.** $\frac{1}{2}[\sec x\tan x + \log(\sec x + \tan x)]$.
37. $\frac{1}{13}e^{2x}(3\sin 3x + 2\cos 3x)$. **49.** $e^x/(1 + x)$.
50. $-e^{-x}(x^2 + 2x + 2)$. **51.** .584. **54.** $\pi/2$. **55.** 11.18. **56.** .429.

EXERCISE LXXXVIII (PAGE 131)

1. $\frac{1}{2}(\text{arc}\sin x)^2$.
5. $1/4\cos^4 x - 1/2\cos^2 x$.
9. $-3x^2\cos x/3 + 18x\sin x/3 + 54\cos x/3$.
10. $38/3$.
13. $2\sqrt{x^2+5x}$.
17. $\pi/4$.
20. $3\,\text{arc}\sin\frac{1}{2}\log x$.
22. $\frac{1}{6}$.
26. $\frac{2}{3}\left\{\sqrt{x^3+1} + \frac{1}{2}\log\left[\left(\sqrt{x^3+1}-1\right)/\left(\sqrt{x^3+1}+1\right)\right]\right\}$.
31. $\frac{2}{3}(\text{arc}\tan x)^{3/2}$.
34. $e^{\sin^2 x}$.
35. $\frac{1}{3}[x - \frac{1}{3}x^3 - (1-x^2)^{3/2}\sin^{-1}x]$.
45. $\frac{1}{6}\log(3x^2 - 2\cos 3x)$.
48. $54\sqrt{x} - 18x^{3/2} + \frac{18}{5}x^{5/2} - \frac{2}{7}x^{7/2}$.
51. $\frac{2}{3}\sqrt{x^3+1}$.
53. $\pi/16$.
56. $\cos 1/x$; $\pi/3$.
58. $\log(x+1) + 2\,\text{arc}\tan\sqrt{x}$.
65. $-2/x - \frac{3}{2}\log(x^2+3)$.
67. $x\,\text{arc}\tan x - \log\sqrt{1+x^2} - \frac{1}{2}(\text{arc}\tan x)^2$.
70. $-4\sqrt{1+2x-3x^2}/x + 4\log\left[\left(x+1+\sqrt{1+2x+3x^2}\right)/x\right]$.
71. $(x^2+1)\,\text{arc}\tan x - x$.
72. $a^x b^{2x} e^{3x}/(\log ab^2 + 3)$.
76. $\frac{1}{2}x^2 - 5x + \frac{1}{2}\log[(x+3)^{17}/(x-1)]$.
80. $\frac{1}{16}(2x - \sin 2x)$.
82. $\frac{1}{3}x^3\log 2x - \frac{1}{9}x^3$.
83. $[x^2a^x - 2xa^x/\log a + 2a^x/(\log a)^2]/\log a$.
84. $\frac{1}{5}x^5\log^2 x - \frac{2}{25}x^5\log x + \frac{2}{125}x^5$.

EXERCISE LXXXIX (PAGE 133)

1. $x + \frac{1}{8}x^4 - \frac{1}{56}x^7 \cdots$.
2. $x + \frac{1}{9}x^3 - \frac{1}{45}x^5 + \cdots$
7. $\frac{3}{4}x^{\frac{4}{3}} + \frac{3}{14}x^{\frac{7}{3}} - \frac{3}{80}x^{\frac{10}{3}} + \cdots$.
11. $x - \dfrac{x^3}{3} + \dfrac{x^5}{10} + \cdots$.
12. $\frac{2}{3}x^{\frac{3}{2}} + \frac{2}{5}x^{\frac{5}{2}} + \frac{1}{7}x^{\frac{7}{2}} + \cdots$.
13. $\log x + 1/2\cdot 3!\ x^2 - 1/4\cdot 5!\ x^4 + \cdots$.
17. $x - x^3/12 - x^5/480 + \cdots$.
22. $\tan\theta - \frac{1}{3}\tan^3\theta + \frac{1}{5}\tan^5\theta + \cdots$.
23. $\sin x + \frac{1}{6}\sin^3 x + \frac{3}{40}\sin^5 x + \cdots$.

EXERCISE XC (PAGE 133)

1. (a) e^x; (b) $x^{e+1}/(e+1)$.
2. (a) $\log x + x + x^2/4 + \cdots$; (b) $-e^{1/x}$.
3. (a) $\frac{1}{2}e^{x^2}$; (b) $e^x(x^2 - 2x + 2)$.
5. (a) $\frac{2}{5}(1+x)^{5/2} - \frac{2}{3}(1+x)^{3/2}$; (b) $\frac{2}{3}x^{3/2} + \frac{2}{5}x^{5/2}$.
7. (a) $\sqrt{x^2-4}$; (b) $\sqrt{x^2-4} - 2\,\text{arc}\sec x/2$.
9. (a) $\sin e^x$; (b) $-x\cos x + \sin x$.
10. (a) $\frac{1}{2}\tan^2 x$; (b) $\frac{1}{2}[\sec x\tan x - \log(\sec x + \tan x)]$.
17. (a) $\log(\log x)$; (b) $\frac{1}{2}\log^2 x$; (c) $\frac{1}{4}x^2(2\log x - 1)$.

EXERCISE XCI (Page 135)

1. $\frac{2}{45}(3x^3-2)(1+x^3)^{3/2}$.
3. $x^2/2\sqrt{1-x^4}$.
4. $\frac{1}{4}x^2\sqrt{x^4+4}-\log(x^2+\sqrt{x^4+4})$.
5. $\frac{1}{4}[2\sqrt{1+x^4}+\log(\sqrt{1+x^4}-1)/(\sqrt{1+x^4}+1)]$. (The reduction formula fails in example 5, and it is best to rationalize.)
6. $\frac{1}{40}(5x^6-6x^3+9)(1+x^3)^{2/3}$.
7. $\frac{1}{16}[\arcsin x^2-\sqrt{1-x^4}(x^2-2x^6)]$.
8. $\frac{1}{4}[x^2\sqrt{1+x^4}+\log(x^2+\sqrt{1+x^4})]$.
9. Final integration by rationalization.
14. $-\frac{1}{10}(2x^5+1)/(1+x^5)^2$.
17. $\frac{1}{32}[(5x^4+2x^{12})\sqrt{1+x^8}+3\log(x^4+\sqrt{1+x^8})]$.
18. $-\frac{1}{12}(1+x^8)^{3/2}/x^{12}$.
20. $\frac{5}{192}(1+x^8)^{\frac{3}{5}}(3x^8-5)$.
19. $\frac{1}{12}(3x^4+2x^{12})/(1+x^8)^{3/2}$.
24. $-\frac{1}{12}(4-x^2)^{3/2}/x^3$.

EXERCISE XCII (Page 136)

1. $\log[(\sqrt{1-3x}-1)/(\sqrt{1-3x}+1)]$.
2. $x/2-\sin x/2$.
4. $\frac{1}{13}(3e^{2x}\sin 3x+2e^{2x}\cos 3x)$.
3. $\frac{12}{37}e^{x/2}(\frac{1}{6}\sin 3x-\cos 3x)$.
9. $\frac{1}{25}[\log(4-5x)+4/(4-5x)]$.
16. $2\sqrt{x^2-a^2}+3\log(x+\sqrt{x^2-a^2})$.
18. $-\frac{1}{2}x\cos 2x+\frac{1}{4}\sin 2x$.
19. $2\sqrt{1-x}+\log[(\sqrt{1-x}-1)/(\sqrt{1-x}+1)]$.
23. $-\frac{1}{3}(10-2x)\sqrt{x+1}$.
25. $\sqrt{x^2+9}+2\log(x+\sqrt{x^2+9})$.
29. $-\frac{1}{2}[\cos^5 x/\sin^2 x+\cos^3 x+3\cos x+3\log(\csc x-\cot x)]$.
31. $\frac{1}{3}x^3(\log^2 x-\frac{2}{3}\log x+\frac{2}{9})$.
42. $\arcsin\frac{1}{2}(e^x-1)$.
35. $-\frac{1}{4}e^{-2x}(2x^2+2x+1)$.
46. $2\sqrt{\sin t}-\frac{2}{5}(\sin t)^{5/2}$.
39. $\frac{1}{3}\sin^3 x-\frac{1}{5}\sin^5 x$.
47. $\frac{1}{3}\arcsin\sqrt{2}\,x^{3/2}$.
48. $\frac{1}{2}[\sqrt{(4+x^2)(3+x^2)}+\log(\sqrt{4+x^2}+\sqrt{3+x^2})]$.
49. $-\sqrt{(1+\sin x)(2-\sin x)}-3\arcsin\sqrt{(2-\sin x)/3}$.
50. $-(2ax^2-x^4)^{3/2}/6ax^6$.
52. $2[\log(2+\sin t)+2/(2+\sin t)]$.
51. $e^x/2\sqrt{4e^x-e^{2x}}$.
53. $\frac{1}{5}(1+e^{-x})(3-2e^{-x})^{3/2}$.
56. $-\sqrt{1+2e^x-e^{2x}}-\arcsin[(e^x-1)/\sqrt{2}]$.
65. $-\frac{1}{2}[(e^{-x}-2)\sqrt{4e^{-x}-e^{-2x}}+4\,\mathrm{vers}^{-1}\frac{1}{2}e^{-x}]$.

EXERCISE XCIII (Page 138)

1. $183\frac{3}{4}+96\log 2$.
6. $2\frac{2}{15}$.
10. $\frac{1}{5}\log\frac{3}{2}$.
18. 1.
2. $7\frac{1}{15}$.
7. $\frac{1}{3}+\frac{1}{6}\sqrt{3}$.
11. $\frac{1}{2}\log\frac{5}{3}$.
19. $2\sqrt{3}/3-\pi/6$.
4. $6+\frac{5}{2}\log 5$.
8. $\pi/3\sqrt{3}$.
12. $2+2\log\frac{5}{3}$.
20. $-3/5+\log 2$.
5. $10+\frac{9}{2}\log 3$.
9. $1\frac{4}{15}$.
17. $\frac{1}{2}(e^\pi+1)$.

EXERCISE XCIV (PAGE 139)

1. $k \log(v_2/v_1)$.
2. $w = k \log(v_2 - b)/(v_1 - b) - a(1/v_1 - 1/v_2)$.
3. (a) 160 ft.; (b) 1280 ft.; 256 ft./sec., average.
4. 570.
5. 63 units.
6. 579.75 large calories.
7. 1/8.
8. $\frac{1}{2}kRPi_0^2$.
9. $\pi E_0 i_0/k \cdot \cos(\phi - \theta)$.
10. (a) 455.4 sec. (b) 23.4 sec. (c) 133.4 sec.
11. 8.96 min.
12. (a) $14\pi r^{5/2}/(15\mu f\sqrt{2g})$; (b) 11.2 min.; (c) 4.44 min.
13. (a) $\frac{2}{3}\mu a\sqrt{2g}[(h+b)^{3/2} - h^{3/2}]$; (b) 608 cu. ft./sec.
15. 9.88 ft.
16. (a) $t = 1/k \cdot \log[a/(a-y)]$; (b) ∞; (c) about $55\frac{1}{2}\%$; (d) 106, 151, 256 min. respectively.
17. (a) $t = 1/k(b-a) \log[a(b-y)/b(a-y)]$; (b) $y = ab(e^{akt_1} - e^{bkt_1})/(ae^{akt_1} - be^{bkt_1})$; (c) $t = y/ak(a-y)$; (d) $1/ak$.
18. (a) $t = 4/a \cdot \log[T_1^{1/4}(1 + bT_2^{\frac{1}{4}})/T_2^{1/4}(1 + bT_1^{1/4})]$; (b) 332 sec.
19. (a) $v = \sqrt{2gR^2s/h(h-s)}$; $v = \sqrt{v_0^2 + 2gR^2s/h(h-s)}$; (b) $\sqrt{gR}$; (c) $\sqrt{v_0^2 + gR}$; (d) $v = \sqrt{v_0^2 + 2gR(h-R)/h}$; $v = \sqrt{v_0^2 + 2gR}$.
20. (a) $t = 1/R \cdot \sqrt{h/2g}\,[\sqrt{h(h-s)} + h \operatorname{arc\,sin} \sqrt{s/h}]$; (b) $t = \sqrt{h/(2gR^2 - v_0^2h)} \cdot [(h-s)(k+s) + (h+k) \operatorname{arc\,sin} \sqrt{(h-s)/(k+s)}]$, where $k = v_0^2h^2/(2gR^2 - v_0^2h)$.
21. (a) .55 + sec.; (b) 9.42 + ft./sec.; (c) $t = 1/2c\sqrt{g} \cdot \log[(\sqrt{g} + cv)/(\sqrt{g} - cv)]$; (d) $v = \sqrt{g}/c \cdot \tanh c\sqrt{g}t$ (tanh = hyperbolic tangent); (e) $s = 1/2c^2 \cdot \log[g/(g - c^2v^2)]$; (f) $\sqrt{g}/c$.
22. (a) $1/c\sqrt{g} \cdot \operatorname{arc\,tan} cv_0/\sqrt{g}$; (b) 2.398 sec., or .727 sec. sooner than when there is no air resistance.
24. (a) 104.17 ft.; (b) $v = 68.5$ ft./sec., $t = 2.715$ sec.

EXERCISE XCV (PAGE 144)

1. (a) $5\frac{1}{3}$; (b) $13\frac{2}{3}$; (c) $20\frac{5}{6}$; (d) 18; (e) 7/6; (f) 9; (g) $20\frac{1}{4}$; (h) $15\frac{3}{4}$; (i) 27/16; (j) $81\frac{1}{15}$; (k) 64/15; (l) 6; (m) 2/5; (n) 1/3; (o) 4/15; (p) 4/15; (q) $18(e^{\pi/6} + 1)/37$; (r) $2(e^{\pi/6} - 1)/37$; (s) $21\frac{1}{3}$; (t) $4\sqrt{2} - 2$.
2. (a) 2; (b) 8; (c) $4/\pi$; (d) $6/\pi$; (e) $\pi/2$; (f) $\pi/6$; (g) 4; (h) 8.
3. (a) 2; (b) 3; (c) 4; (d) 3/2; (e) $\pi/(4 - \pi)$; (f) 8.4 +; (g) 5; (h) $(32 - 3\pi)/3\pi$; (i) 1.84; (j) 1.71; (k) $2/(\pi - 2)$; (l) $(\pi - 2\log 2)/2\log 2$.
4. (a), (b), (c) $4\frac{1}{2}$; (d) $1\frac{1}{3}$; (e) $10\frac{2}{3}$; (f) 1/3.

EXERCISE XCVI (PAGE 145)

1. (a) $19/27$; (b) $\frac{2}{3}(5\sqrt{5}-1)=6.78$; (c) $2\pi/3$; (d) $\frac{1}{9}(7\sqrt{7}-1)$;
(e) $6+\frac{5}{2}\log 5$; (f) $\frac{10}{3}+\frac{3}{2}\log 3$; (g) 1.081;
(h) $1+\sqrt{8}\log[\sqrt{8}(3+\sqrt{8})/(4+\sqrt{8})]$; (i) $\log(2+\sqrt{3})$.
2. (a) $\frac{1}{9}(512-7\sqrt{7})$; (b) 4.56; (c) $(e-e^{-1})=2.35$.

EXERCISE XCVII (PAGE 146)

1. (a) $\pi/7$; (b) $156\frac{1}{4}\pi$; (c) $\frac{1}{2}\pi^2$; (d) $\pi^2/4$; (e) $34\frac{2}{15}\pi$;
(f) $(e^{2ab}-1)\pi/2a$.
2. (a) $8\pi/3$; (b) $32\pi/15$; (c) $(64\sqrt{2}-32)\pi/3$; (d) 4π.
3. (a) $512\pi/15$; (b) $512\pi/15$; (c) $64\pi/3$; (d) $729\pi/35$.
4. (a), (b), (c) $16\pi/15$; (d) $2048\pi/35$; (e) $2\pi(57-80\log 2)$;
(f) $2\pi(3\sqrt{3}-2\pi/3)$; (g) $\pi\sqrt{2}/60$; (h), (i) $81\pi\sqrt{2}/20$;
(j) $92\pi/15\sqrt{5}$; (k) $2\sqrt{2}\pi/15$; (l) $\pi(57-80\log 2)/\sqrt{2}$.
5. (a) 9π; (b) $72\pi/5$; (c) $56\pi/15$; (d) $128\pi/3$; (e) $34\pi/15$; (f) $104\pi/35$;
(g) 18π; (h) $32\pi/3$; (i) $5\pi/14$; (j) $2\pi/5$; (k) $3\pi/10$; (l) $3\pi/10$.
6. (a) $\pi^2/4$; (b) $.4674\pi$.
7. (a) $256/15$; (b) 61.28.
8. $\pi/2a$.
9. V about $OX=10\pi$, about $OY=176\pi/3$.
10. $1824\pi/15$.

EXERCISE XCVIII (PAGE 148)

1. (a) $98\pi/81$; (b) $56\pi/3$; (c) 49π; (d) $56\pi/3$; (e) 4π;
(f) $(820-31\log 3)\pi/72=31.89$.
2. 14.42.
3. $\frac{8}{3}\pi(17\sqrt{17}+2\sqrt{2}-2)=594.1$.

EXERCISE XCIX (PAGE 148)

1. (a) 4; (b) 9; (c) $\pi/2$; (d) $\pi/4$; (e) $3\pi/2$; (f) $3\pi/2$; (g) $19\pi/2$;
(h) $9\pi/2$; (i) $3\pi/4$; (j) $3\pi/4$; (k) $3\pi/2$; (l) $3\pi/2$; (m) $9\pi/2$;
(n) $9\pi/2$.
2. (a) $(4\pi+3\sqrt{3})/(2\pi-3\sqrt{3})$; (b) $(10\pi+9\sqrt{3})/(5\pi-9\sqrt{3})$;
(c) $(4\pi+3\sqrt{3})/(2\pi-3\sqrt{3})$.
3. (a) $\frac{1}{4}(\pi-1)$; (b) $\frac{1}{2}(1-1/\sqrt{2})$; (c) $5\pi/4-2$;
(d) the three areas are $(\pi/2-1)$, $(1-\pi/4)$, and $(\pi+1)$; (e) $\frac{5}{8}\pi-\sqrt{3}$;
(f) $\frac{3}{4}(\pi-\sqrt{3})$; (g) $\frac{1}{6}(6+\pi-3\sqrt{3})$; (h) $\frac{1}{4}(\pi+9-3\sqrt{3})$;
(i) $\frac{1}{12}\pi+\frac{1}{4}+5\sqrt{3}/32$.
4. (a) 4; (b) $\frac{3}{2}\pi$; (c) $\frac{16}{3}$; (d) 8.
5. (a) $2[\sqrt{2}+\log(\sqrt{2}+1)]$; (b) $\frac{8}{3}$.
6. (a) 16; (b) $128\pi/15$.

EXERCISE C (PAGE 150)

1. (a) $\frac{1}{4}\pi$; (b) $\frac{1}{12}\pi$; (c) $\frac{4}{3}\pi$; (d) $\frac{14}{15}\pi$; (e) $\frac{2}{9}\pi$; (f) $\frac{2}{9}\pi$.
2. (a) $\frac{32}{3}$; (b) $\frac{16}{3}$; (c) $8\sqrt{3}/3$; (d) $32\sqrt{3}/3$; (e) $4\sqrt{3}$.
3. 6π.
4. $125\pi/2$.
5. $\frac{8}{3}\pi$.
6. (a) $341\frac{1}{3}$; (b) $512\pi/3$.
7. $13\frac{3}{5}$.
8. $\frac{8}{3}a^3$.

EXERCISE CI (PAGE 152)

1. 1.110.
2. 1.093.
3. 1.070.
4. 1.088.
5. .100.
6. .100.
7. .946.
8. .9045

EXERCISE CII (PAGE 152)

1. $A_T = 6.414$, $A_S = 6.276$.
2. $A_T = 4.75$, $A_S = 4.69$.
3. 1.07.
4. 1.09.
5. $A_T = 7.385$, $A_S = 7.364$.
7. 6.11.
8. 4.12.
11. .85.
12. .90.
13. $A_T = 1.911$, $A_S = 1.914$.
15. 1.057.
16. .737.
17. 1.16.
18. 1.18.
19. .86.
20. .42.
23. 2.598.
27. $\pi = 3.14159265$.

EXERCISE CIII (PAGE 154)

(S = curved surface)

1. $A = 3\frac{5}{9}$, $P = 12.256$, $V = 256\pi/45$, $S = 30.464$.
2. $A = 32\sqrt{5}/3$, $P = 22.967$, $V = 160\pi$, $S = 296.3$.
3. $A = 1.913$, $P = 5.779$, $V = 11\pi/3$, $S = 4\pi$.
5. $P = 18.611$, $A = 4$, $V = 128\pi/7$, $S = 203.05$.
6. $A = 22.36$, $P = 28.59$, $V = 490.875$, $S = 133.1$ (134.4 by trapezoidal rule).
7. $A = 3.86$, $P = 2.914$, $V = .592$, $S = 2.864$.
8. $A = 1$, $P = 4.483$, $V = 2.466$, $S = 7.225$.
9. $A = 1.718$, $P = 6.722$, $V = 10.036$, $S = 26.276$.
11. $A = 1/6$, $P = 3.651$, $V = .209$.
13. $A = 3\pi/2$, $P = 8$, $V = 8\pi/3$, $S = 32\pi/5$.
14. $A = 3\pi$, $P = 8 + 2\pi$, $V = 5\pi^2$, $S = \frac{64}{3}\pi$, $V_t = 7\pi^2$.

EXERCISE CIV (PAGE 156)

1. $-\frac{10}{3}$.
2. 2.
3. 1.
4. 9.
5. $2\frac{1}{4}$.
6. $\frac{3}{4}$.
7. $18\frac{9}{20}$.
8. $-\frac{64}{21}$.
9. $\frac{11}{84}$.
10. $\frac{13}{60}$.
11. $\frac{101}{84}$.
12. $\frac{67}{60}$.
13. $(4 + 3\pi)/12$.
14. $\frac{17}{20}$.
15. $\frac{4}{35}$.
16. $\frac{11}{60}$.

EXERCISE CV (Page 157)

1. $4\frac{1}{2}$. 2. $\frac{4}{3}$. 3. $4\frac{1}{2}$. 4. $4\frac{1}{2}$. 6. $4\frac{1}{2}$. 7. $7\frac{1}{2} - 4\log 4$. 9. 8. 10. $13\frac{1}{3}$. 11. $5\frac{1}{3}$. 12. 9. 13. 1. 14. 3. 15. $4\frac{1}{2}$. 16. $21\frac{1}{3}$. 17. 16. 18. 4. 19. 2. 20. 2. 21. 2. 22. $7\frac{1}{2}$. 23. 6. 24. 16. 25. 4. 26. $1\frac{1}{2}$. 30. $10\frac{2}{3}$. 31. .414. 32. 19.2.

33. (a) $(8\pi + 3\sqrt{3})/(4\pi - 3\sqrt{3})$; (b) $(8\pi + 3\sqrt{3})/(4\pi - 3\sqrt{3})$, or a little more than 4 : 1; (f) 9 : 1; (g) 24.3/13.4; (h) nearly 5.

EXERCISE CVI (Page 158)

1. $15\pi/4$.
2. $\frac{3}{4}(\pi - 3\sqrt{3}/4)$.
3. $\frac{3}{4}(\pi + 3\sqrt{3}/2)$.
4. 2π.
5. $7\pi/36$.
6. $\pi/12 + \sqrt{3}/8$.
7. $3\pi/2 + 9\sqrt{3}/16$.
8. $\pi/12 - \sqrt{3}/16$.
9. $\pi/2 - 9\sqrt{3}/16$.
10. $\pi/4 + 2$.
11. $9\sqrt{3}/8 - \pi/4$.
12. $5\pi/6 + 7\sqrt{3}/8$.
13. $7\sqrt{3}/8 - 5\pi/12$.
14. $2 - \pi/4$.
15. $9\sqrt{3}/8 + \pi/2$.
16. π.
17. $\pi/2 - 2/3$.
18. $8\pi/3 - \sqrt{3}$.
19. $\frac{1}{27}(8\pi + \sqrt{3})$.
20. $3\pi/4 + 4/3$.
22. 1/2.
23. $\frac{1}{2}(4 - \pi)$.
24. .3424.
25. 1.027.
26. $.2392\,a^2$.
27. 5.504.
28. $2.457\,a^2$.

EXERCISE CVII (Page 160)

1. $\frac{1}{6}abc$. 2. 9. 3. $5\frac{1}{3}$. 4. $\frac{4}{15}$. 5. $\frac{17}{20}$. 6. $\frac{5}{12}$. 7. $\frac{1}{2}$. 8. $\frac{2}{3}$. 9. $143\sqrt{2}/15$. 10. 12π. 11. 4π. 12. 4π. 13. $\frac{4}{3}\pi r^3$. 14. 60.62 cu. in.

EXERCISE CVIII (Page 162)

1. (a) $6\,W$; (b) $2\,W$; (c) $4\,W$; (d) $\frac{9}{2}\,W$; (e) $\frac{16}{3}\,W$; (f) $\frac{22}{3}\,W$; (g) $6\,W$; (h) $5\,W$; (i) $2\,W$; (j) $12\,W$; (k) $\frac{26}{3}\,W$.
2. $\frac{128}{3}\,W$.
3. (a) $8\,W$; (b) $12\,W$.
4. (a) $4\,W$; (b) $\frac{16}{3}\,W$; (c) $\frac{128}{3}\,W$; (d) $\frac{256}{15}\,W$.
5. (a) $\frac{4}{3}\,W$; (b) $\frac{8}{15}\,W$.
6. (a) $1920\,W$; (b) $1280\,W$.
7. $266\frac{2}{3}$ lb.
8. $5290\frac{2}{3}$ lb.; $10{,}581\frac{1}{3}$ lb.
9. $7440\,\pi$ lb.
10. 14,880 lb.
11. $5952\,\pi$ lb.
12. (a) 4036 lb.; (b) 8711 lb.

EXERCISE CIX (Page 164)

1. $\frac{1}{2}kab(2\,l + b\sin\theta)$.
2. 7885 lb.
3. (a) $\frac{1}{2}\pi k a^2(2\,c \pm l\cos\theta)$; (b) $372\,\pi$ lb. and $1612\,\pi$ lb.
4. $360\,\pi k$.
5. 2058 lb.; 1920 lb.; 696 lb.
6. $\pi k v^2 a h/2\,g$.
7. 19,100 lb.

8. (a) $\frac{1}{2}\pi a^2h^2k$ ft.-lb.; (b) $156{,}250\pi/3$ ft.-lb.
9. $108{,}000\,\pi$ ft.-lb.
10. 248.5 ft.-tons.
11. $3\pi a^3k/16$.
12. $\pi^4k^2(8a + 3\pi k)/192$.
14. (a) $32(\pi - 2)$; (b) 64.
15. $4mr^2$.
16. $\frac{26}{3}\pi a^2$.
17. $\frac{1}{2}F_0\pi a^3$.
18. $\pi F_0(a^4 - b^4)/2a$.
19. 41 : 15.
20. $kMm/a(a + l)$.
21. $2kmM/\pi a^2$.
22. $2kMm/a^2 \cdot \left(1 - h/\sqrt{h^2 + a^2}\right)$.
23. 349.5 tons.
24. $2\pi kI\left(1 - h/\sqrt{h^2 + a^2}\right)$.
25. (a) Light on path is about 24% as much as that on the circle; (b) about 22%.
26. $10\,\pi$ kilograms.
27. 26.4+ kilograms.
28. 20 kg.
29. 16.07 kg.
30. 17.5 kg.
31. $1250\,\pi$ kg., or 3927 kg.
32. 3590 kg.
33. $F = 22.05$ kg., $P = 3919$ kg., $F/P = .0056$.
34. $\frac{4}{3}\mu p\pi^2a^3$.
35. (a) 7369 kg.; (b) 7106 kg.
36. $2\mu pa^3\pi^2(\alpha - \sin\alpha\cos\alpha)/\sin^3\alpha$.
37. πpa^2 kg.
38. $\frac{4}{3}\pi^2\mu pa^2\sqrt{a^2 + h^2}$ or $\frac{4}{3}\pi^2\mu pa^3/\cos\alpha$, where α is the half angle at the vertex.
39. 12,800 kg.-cm.; 6032 kg.
40. (a) 24.1; (b) nearly 8 kg. per sq. cm.; (c) 8.63 kg. per sq. cm.
41. (a) $2\pi^2\mu kr^2(1 - \cos 2\alpha)$; (b) $p_a = 2.48$, $W = 346.7$ kg.-cm.
42. (a) Front, 220,560.8 lb.; back, 179,439.2 lb.
(b) $S_0 = 5233.64$ lb. per foot of width of wall; $S_1 = 12{,}710.28$ lb.; $S_2 = 20{,}186.92$ lb.; $S_3 = 23{,}925.24$.
(c) $r = 373.832$ lb./ft.^{2}

EXERCISE CX (Page 170)

1. $(\frac{3}{5}, -\frac{1}{2})$.
2. $(6\frac{4}{5}, -4)$.
3. (2, 0).
4. $(2\frac{1}{5}, 0)$.
5. $(\frac{1}{2}, \frac{3}{10})$.
6. (1, 1).
7. $(1, \frac{4}{5})$.
8. $(2, \frac{1}{5})$.
9. $(\frac{3}{2}, \frac{7}{10})$.
10. (2, 0).
11. $(-2\frac{1}{10}, \frac{3}{2})$.
12. $(\frac{3}{2}, -\frac{1}{2})$.
13. $(\frac{9}{10}, 2\frac{7}{20})$.
14. $(1\frac{7}{8}, \frac{6}{7})$.
15. $(3\frac{3}{5}, 3\frac{3}{10})$.
16. $(2\frac{14}{15}, 1\frac{13}{15})$.
17. $(\frac{14}{15}, -\frac{11}{15})$.
18. $\left(8\sqrt{2}/15, 8\sqrt{2}/15\right)$.
19. $(\frac{19}{50}, 2\frac{23}{50})$.
20. $(1\frac{2}{5}, -\frac{1}{7})$.
22. (a) (3.08, 2.95); (b) (1.37, 1.30).
23. (a) (3.42, − 3.26); (b) (6.904, .594).
24. (.695, 2.478).
25. (2.20, .71).
26. (a) (.27, 2.18); (b) (−1.16, .107).

EXERCISE CXI (Page 172)

1. $33\frac{1}{3}$.
2. $2\frac{2}{3}$.
3. $a^2/2\pi$, $2a^2/3\pi$.
4. (a) 1.33; (b) 1.64.
5. 2.
6. $2/\pi$.
7. 1/2.
8. 3/8.
9. 3/8.
10. 1/2.
11. (a) 80; (b) 116; (c) 40.
12. (a) $53\frac{1}{3}$; (b) $81\frac{2}{3}$.
13. 40 ft./sec.
14. (a) $2a\omega/\pi$; (b) 16 ft./sec.
15. $170\frac{2}{3}$ ft.^{3}/sec.

16. 675.2 ft.^{3}/sec.; 607,680 cu. ft.
17. $3a/8$.
18. (a) $8a/15$; (b) $5a/8$.
19. 3 in. from the base.
20. $1\frac{1}{2}$ ft. from the base.
21. $v = 4\frac{1}{2}$ cu. ft., $\bar{x} = 1\frac{3}{5}$ ft.
22. $1\frac{1}{3}$ ft. from the bottom.
23. 8 ft. 2 in. below the surface.
24. 4.33 ft. from the base.
25. Three fourths of the way down.
26. $2\frac{2}{7}$ ft.
27. 11,904 lb.; $6\frac{4}{9}$ ft. below the surface.
28. (a) 3.76 ft. below the surface; (b) 1.18 ft. below the surface.
29. About 7 ft. 4 in. below the surface.
30. $(\frac{3}{8}a, \frac{3}{8}a, \frac{3}{8}a)$.
31. Example 4, $(\frac{3}{7}, \frac{5}{16}, \frac{13}{28})$; example 6, $(\frac{9}{25}, \frac{7}{25}, \frac{11}{25})$; example 7, $(\frac{2}{5}, \frac{2}{5}, \frac{2}{3})$.
32. $\bar{x} = .42$, $\bar{y} = 1.06$; $\bar{x} = [3/\sqrt{2} - \log(3 + 2\sqrt{2})][a/\sqrt{2} + \log(1 + \sqrt{2})]$;
$\bar{y} = 4(2\sqrt{2} - 1)a/3[\sqrt{2} + \log(1 + \sqrt{2})]$.

EXERCISE CXII (Page 175)

1. $I_x = 6\frac{2}{5}$, $I_y = 1\frac{1}{7}$, $I_0 = 7\frac{19}{35}$.
2. $I_x = 13/60$, $I_y = 1\frac{17}{84}$, $I_0 = 1\frac{44}{105}$.
3. $I_x = 4096/105$, $I_y = 128/15$.
7. (a) $2\frac{2}{15}$; (b) $(\frac{6}{7}, 1\frac{7}{8})$; (c) $I_y = 128/63$, $I_x = 512/55$.
8. (a) $4\frac{1}{2}$; (b) $(\frac{3}{5}, -\frac{1}{2})$; $I_x = 3\frac{3}{20}$, $I_y = 5\frac{29}{140}$, $I_0 = 8\frac{5}{14}$.
9. Example 1, $r_x^2 = \frac{32}{5}$, $r_y^2 = \frac{6}{7}$, $r_0^2 = 198/35$;
example 2, $r_x^2 = 13/70$, $r_y^2 = 101/98$, $r_0^2 = 298/245$;
example 3, $r_x^2 = \frac{128}{35}$, $r_y^2 = \frac{4}{5}$;
example 7, $r_x^2 = 20/21$, $r_y^2 = 512/55$.

EXERCISE CXIII (Page 176)

1. $8\pi/3 - \sqrt{3}$.
2. $\pi/128 + \frac{1}{16}$.
3. $\frac{1}{8}$.
4. $\frac{5}{3} - \pi/2$.
5. $11\sqrt{3}/8 - 2\pi/3$.
6. $\pi/12 + \sqrt{3}/8$.
7. $\frac{9}{4}(\pi - \sqrt{3}/2)$.
8. $195\pi/32$.

EXERCISE CXIV (Page 178)

1. $cy = e^{\sqrt{1-x^2}}$.
2. $3y + y^3 - 9\log x = c$.
3. $\log c(x + \sqrt{1 - x^2}) = \text{arc} \sin y$.
4. $\sqrt{(1+x)/(1-x)} = c(y + \sqrt{1 + y^2})$.
5. $e^{-x} - \log c(\csc y - \cot y) = 0$.
6. $y/\sqrt{1 - y^2} = (x + c)/(1 - cx)$.
7. $cx = ye^{\frac{x+y}{xy}}$.
8. $\log c\sqrt{1 + x^2} + \text{arc} \tan y = 0$.
9. $y = \pm\sqrt{c^2x^2 - 1}/cx$.
10. $y = \sqrt{\frac{1}{3} \text{arc} \tan 3x/2 + c}$.
11. $y = \sqrt{\log[c(2x - 3)/(2x + 3)]}$.
12. $y = 1 - 1/c\sqrt{x^2 + 1}$.
13. $y = \tan \log c/\sqrt{1 + x^2}$.
14. $ce^{-1/y} = \sqrt{2x + 1}$.
15. $(1 + x^2)(1 + 2y) = c$.
16. $y = -x + \log c(x + 1)^2$.
17. $x = y^2/2 + \log cy$.
18. $cy = \sqrt{1 + e^{2x}}$.

EXERCISE CXV (PAGE 178)

1. $c_1e^{2t} + c_2e^{3t}$.
2. $c_1e^{2t} + c_2e^{-3t}$.
3. $c_1e^{2t} + c_2te^{2t}$.
4. $c_1e^{2t} + c_2e^{-2t}$.
5. $c_1 \sin 2t + c_2 \cos 2t$.
6. $c_1 + c_2e^{4t}$.
7. $c_1e^t + c_2e^{-t}$
8. $c_1e^{(1+\sqrt{2})t} + c_2e^{(1-\sqrt{2})t}$.
9. $e^{2t}(c_1 \cos 3t + c_2 \sin 3t)$.
10. $e^{-2t}(c_1 \cos 2t + c_2 \sin 2t)$.
15. $c_1e^{\frac{4}{3}t} + c_2e^{-\frac{2}{3}t}$.
16. $e^{4t}(c_1 \cos 3t + c_2 \sin 3t)$.
19. $c_1e^{\frac{1+\sqrt{6}}{3}t} + c_2e^{\frac{1-\sqrt{6}}{3}t}$.
20. $e^{-\frac{1}{4}t}(c_1 \cos t/\sqrt{2} + c_2 \sin t/\sqrt{2})$.
23. $c_1 + c_2e^{3t} + c_3e^{-4t}$.
24. $c_1 + c_2t + c_3e^{2t} + c_4e^{-2t}$.
25. $c_1e^t + c_2e^{-t} + c_3 \sin 2t + c_4 \cos 2t$.
28. $c_1e^{-t} + c_2te^{-t} + c_3t^2e^{-t}$.
29. $c_1e^{-t} + c_2te^{-t} + c_3t^2e^{-t} + c_4$.
30. $c_1 + c_2e^{\sqrt{2}t} + c_3e^{-\sqrt{2}t} + c_4 \cos \sqrt{2}t + c_5 \sin \sqrt{2}t$.

EXERCISE CXVI (PAGE 180)

1. $e^{3t}(c_1 \cos 2t + c_2 \sin 2t) + 3$.
2. $e^{2t}(c_1 \cos \sqrt{3}t + c_2 \sin \sqrt{3}t) + 2$.
3. $c_1e^{3t} + c_2e^{-2t} - \frac{2}{3}t + \frac{1}{9}$.
4. $c_1 \cos 3t + c_2 \sin 3t + \frac{1}{9}t + \frac{1}{18}$.
5. $c_1 \cos 3t + c_2 \sin 3t + t^2/9 - 2/81$.
6. $e^{-t}(c_1 \cos \sqrt{3}t + c_2 \sin \sqrt{3}t) + t^2/4 + t + 1$.
7. $e^{-3t}(c_1 \cos \sqrt{3}t + c_2 \sin \sqrt{3}t) + \frac{1}{2}t^2 + \frac{1}{2}t + \frac{1}{6}$.
8. $c_1e^{5t} + c_2te^{5t} + t^2 + t + 1$.
9. $e^{-2t}(c_1 \cos 3t + c_2 \sin 3t) + t^2 + t + 1$.
10. $e^t(c_1 \cos t + c_2 \sin t) + t^2 + 3$.
15. $e^t(c_1 \cos \sqrt{7}t + c_2 \sin \sqrt{7}t) + 4t^2 - 22t + 17$.
16. $c_1 \cos 2t + c_2 \sin 2t + t^3 + 2t^2 + 3t + 4$.
17. $c_1e^{2t} + c_2e^{-t} + te^{2t}$.
18. $c_1e^{4t} + c_2e^{-t} - \frac{1}{5}te^{-t}$.
19. $c_1e^{3t} + c_2e^{-t} - \frac{1}{8}e^{2t}$.
23. $c_1 \cos 3t + c_2 \sin 3t + \frac{1}{2}e^{3t}$.
27. $c_1e^t + c_2te^t + 3t^2e^t$.
28. $c_1 \cos 3t + c_2 \sin 3t + \cos 2t$.
29. $c_1 \cos 3t + c_2 \sin 3t + \frac{1}{2}t \sin 3t$.
30. $c_1e^{3t} + c_2e^{-3t} - \frac{1}{3}\cos 3t$.
31. $c_1 \cos 2t + c_2 \sin 2t - 2 \sin 3t$.
32. $c_1 \cos 2t + c_2 \sin 2t - 2t \cos 2t$.
33. $c_1 \cos t/2 + c_2 \sin t/2 - t \cos t/2$.
38. $c_1e^{3t} + c_2e^{-t} - \frac{32}{65}\sin 2t - \frac{56}{65}\cos 2t$.
39. $e^{3t}(c_1 \cos 2t + c_2 \sin 2t) + 2 \sin t - \cos t$.
40. $e^{-t}(c_1 \cos 2t + c_2 \sin 2t) + 2 \sin t + \cos t$.
41. $e^{-t}(c_1 \cos 2t + c_2 \sin 2t) + 4 \cos 2t + \sin 2t$.

EXERCISE CXVII (PAGE 183)

1. $t^4/12 + c_1t + c_2$.
2. $-\sin 2t$.
3. $c_1e^t + c_2e^{-t}$.
4. $\frac{1}{4}e^{2t} + c_1t + c_2$.
5. $c_1e^{2t} + c_2e^{-2t} - \frac{1}{7}e^{3t} + e^t$.
6. $(2x + 1) = ce^{1/y}$.
7. $c_1 \cos 2t + c_2 \sin 2t - \sin 3t + 2 \cos 3t$.
8. $c_1 \cos 2t + c_2 \sin 2t + \frac{1}{2}e^{2t} + t^2 - 2$.

9. $c_1 \cos 2t + c_2 \sin 2t + \sin 3t + 4.$

10. $c_1 \cos 2t + c_2 \sin 2t + e^{2t} + 4 \sin t/2.$

11. $y = (t + c)^{3/2}.$

12. $c_1 + c_2 t + e^{-3t}(c_3 \cos 2t + c_4 \sin 2t).$

13. $c_1 + c_2 t + c_3 e^{(2+\sqrt{3})t} + c_4 e^{(2-\sqrt{3})t}.$

14. $c_1 \cos t/2 + c_2 \sin t/2 - t \cos t/2.$

15. $e^{-t}(c_1 \cos 3\sqrt{3}t + c_2 \sin 3\sqrt{3}t) + \frac{1}{37} \sin t/2 + \frac{36}{37} \cos t/2.$

16. $c_1 e^t + c_2 te^t + \frac{1}{2} e^{3t} + 3t^2 + 12t + 11.$

17. $y = x^{3/2} + c.$

18. $c_1 e^{3t} + c_2 e^{-2t} + e^{4t} + 2e^{-t}$

19. $c_1 e^{3t} + c_2 e^{-t} + te^{3t} + 4.$

20. $e^{-t}(c_1 \cos 2t + c_2 \sin 2t) + \frac{5}{2} te^{2t} + \frac{11}{8} e^{-t}.$

22. $y = (x + c)^3.$

28. $c_1 e^t + c_2 te^t + t^2 e^t.$

29. $c_1 e^{\sqrt{3}t} + c_2 e^{-\sqrt{3}t} - te^t - e^t.$

30. $c_1 e^{2t} + c_2 e^{-2t} - \frac{1}{5} \sin t - te^t - \frac{2}{3} e^t.$

31. $y^2 + 2xy = c.$

32. $cx = e^{\cos y/x}$

33. $cx = x^2 - y^2.$

35. $(c_1 t + c_2)^2 + 4 = c_1 s^2.$

36. $(c_1 t + c_2)^2 + 1 = c_1 (s + 1)^2.$

37. $s = t^2/3 + c/t.$

38. $s = 1 + c \cos t.$

39. $s(t^2 + 1) = \frac{1}{2} t^2 + \log t.$

40. $s = \frac{1}{2}(t^2 - 1) + ce^{-t^2}.$

41. $t^4/2s^2 + \frac{1}{3} t^3 = c.$

42. $s = ct(1 + t) - t.$

43. $x^2 = cy^2 e^{3/xy}.$

44. $y - \log[(x + y)^2 + 2(x + y) + 5] + \frac{3}{2} \arc\tan \frac{1}{2}(x + y + 1) = c.$

EXERCISE CXVIII (PAGE 185)

1. $s = 2e^{-t}(2 \cos 2t + \sin 2t).$

2. $s = \cos 2t + 2t^2 - 1.$

3. $s = 2 \sin 2t + \sin t.$

4. $s = \sin 2t + 2 \cos 2t + t \sin 2t.$

5. (a) $s = 16t^2 + 24t$; (b) $1\frac{1}{2}$ sec.

6. $s = -16t^2 + 96t$; rises 144 ft.

7. $\theta = \alpha \cos 4t$; $P = \frac{1}{2}\pi$ sec.

8. $s = 4 \sin 4t.$

9. $s = 8e^{-2t} \sin 2t.$

10. $C = c_1 \cos 500t + c_2 \sin 500t$; $N =$ nearly 80 cycles per sec.

11. $\theta = -2t^2 + 8\pi t$; 2π sec., 4π rev.

12. $\theta = t^2$; $t = 52.36$ sec., 509 rev.

13. (a) $s = e^{-\frac{k}{2}t}(c_1 \cos \sqrt{4a - k^2}t + c_2 \sin \sqrt{4a - k^2}t)$;

(b) $s = 4e^{-\frac{1}{4}t} \cos \frac{5}{2} t.$

14. $V = V_0 e^{-\frac{R}{2L}t} \cos \sqrt{\frac{1}{KL} - \frac{R^2}{4L^2}}\, t.$

15. (a) $\theta = e^{-\frac{k_1}{2I}t}\left(c_1 \cos \sqrt{\frac{k_2}{I} - \frac{k_1^2}{4I^2}}\, t + c_2 \sin \sqrt{\frac{k_2}{I} - \frac{k_1^2}{4I^2}}\, t\right)$;

(b) $\theta = \alpha e^{-\frac{k_1}{2I}t} \cos \sqrt{\frac{k_2}{I} - \frac{k_1^2}{4I^2}}\, t.$

16. (a) $\theta = c_1 e^{-\left(\frac{k_1 + \sqrt{k_1^2 - 4Ik_2}}{2I}\right)t} + c_2 e^{-\left(\frac{k_1 - \sqrt{k_1^2 - 4Ik_2}}{2I}\right)t}.$

17. (a) $Q = e^{-\frac{R}{2L}t}\left[c_1 \cos\sqrt{\frac{1}{KL} - \frac{R^2}{4L^2}}\,t + c_2 \sin\sqrt{\frac{1}{KL} - \frac{R^2}{4L^2}}\,t\right] + KV;$

(b) $Q = KV\left(1 - \cos\sqrt{\frac{1}{KL} - \frac{R^2}{4L^2}}\,t\right).$

18. (a) $Q = e^{-\frac{R}{2L}t}(c_1 \cos k_1 t + c_2 \sin k_1 t) + kKt - RkK^2$, where

$$k_1 = \sqrt{\frac{1}{KL} - \frac{R^2}{4L^2}}.$$

19. (a) $y = \frac{1}{2} klx^2 - \frac{1}{6} kx^3$; (b) $y = \frac{1}{12} k(6l^2x^2 - 4lx^3 + x^4)$;
(c) $y = \frac{1}{48} k(3l^2x^2 - 2x^4)$; (d) $y = \frac{1}{6} k(3ax^2 + bx^3)$.

INDEX

(The numbers refer to the pages)

ANNOUNCEMENTS

ADVANCED CALCULUS

By EDWIN BIDWELL WILSON, Professor of Mathematics in the Massachusetts Institute of Technology

8vo, cloth, ix + 566 pages, $5.00

CHAPTERS I–X. 8vo, cloth, ix + 280 pages, $2.75

PROFESSOR WILSON'S "Advanced Calculus" is an up-to-date, practical textbook and laboratory manual of higher calculus, providing in a single volume a second course in calculus broad enough to meet the varying needs and attainments of students and the divergent viewpoints of teachers. The material is arranged to give the maximum flexibility and adaptability.

Throughout the work attention has been paid to the needs of students of engineering and mathematical physics; and the exercises, of which a large number have been provided, aim to cultivate in the student a large reliance on himself.

Two chapters in review supply the connection with elementary texts. Then follow four chapters on differential calculus (Taylor's Formula, partial differentiation of explicit and of implicit functions, complex numbers and vectors), four chapters on the integration of differential equations, five chapters on integral calculus, and five on the theory of functions. The first ten chapters are published also in separate form.

125½

GINN AND COMPANY PUBLISHERS

Made in the USA
Monee, IL
13 September 2026

62976468R00146